WE TOOK THE RISK

WE TOOK THE RISK

Mint = Risk
Thanks for all
your support
+ leadership!
Love,

TOM WEIRICH

NEW DEGREE PRESS

WE TOOK THE RISK

ISBN 979-8-88504-639-8 *Paperback*
979-8-88504-957-3 *Kindle Ebook*
979-8-88504-844-6 *Ebook*

To the original risk-takers in my life, my parents—Paul and Jarmila Weirich—who escaped from communist Czechoslovakia in the 1960s to give my sisters and me a better life in the United States. Your hardships, struggles, and successes have shown us that the American Dream is alive and well. You forever continue to challenge me to give back to a country that has given us so much.

Contents

"You cannot get through a single day without having an impact on the world around you. What you do makes a difference, and you have to decide what kind of difference you want to make."

—JANE GOODALL, PRIMATOLOGIST, ANTHROPOLOGIST, AND ACTIVIST

Foreword

Sir Robert Swan, OBE

Leadership to me is the power to be brave and ignite action within yourself and others. A primary tenet of leadership involves risk-taking and providing the inspiration for others to follow you in that risk.

As the first person to have walked to both the North and South Poles, I have seen first-hand the effects of climate change. What I saw reaffirmed a dedication in me toward my life's mission to protect Antarctica and our planet at large. This dedication led to setting up the 2041 Foundation with a mission to engage businesses and communities on climate science, personal leadership, and the promotion of sustainable practices. The main mission is to preserve Antarctica as the last great wilderness on our planet.

I learned a few life lessons as a risk-taker on my expeditions. The first lesson was that great feats are rarely achieved by individuals in isolation; more often, they are a team effort.

Strategic decision-making and committing oneself to completing the mission, even after a deviation from the initial trajectory, is paramount. The second lesson I've learned concerns commitment. Specifically, if you say you're going to do it, do it. There is no turning back.

We as society are now well over halfway into our fifty-year mission of maintaining Antarctica as a natural reserve for science and peace. In the year 2041, the fate of Antarctica will be decided, and I remain committed to ensuring that for the next eighteen years we have the sense to leave this one place on Earth untouched. I believe renewable energy has a major role to play in order to fulfill that mission.

Just as vital as the technology itself is the need to foster and inspire the next generation of risk-taking innovators and entrepreneurs. They truly are the "renewable power" behind sustainable energy.

I believed we had to set an example and do something no one had done before. I organized the first polar expedition powered entirely off of renewable energy in March of 2008. Our small team lived off the power of renewables and sent broadcasts from 2041's Antarctic E-Base via the internet for two weeks.

As we shared the story with students and teachers globally, for the first time in history, a team had attempted to survive in Antarctica relying solely on renewable energy. In doing so, we needed to push the technology for us to survive. In that same vein, we all must continue to push these renewable technologies for us to survive here on Earth.

Tom and I had the privilege of getting to know each other one night in 2010 during ACORE's Leadership Council Dinner in New York through a mutual connection, Michael Naylor. Later in March of 2011, we embarked on an expedition with a handful of renewable energy leaders to experience the impact of climate change on Antarctica firsthand. From day one, I was impressed by his ability to both connect with the other expedition team members and his ability to quickly forge deep life-long connections.

Be it on our hikes avoiding deathly deep crevasses or in post-expedition merriment onboard the ship after a long day exploring the Antarctic Peninsula, Tom always had a genuine interest in hearing people's stories—especially those related to their start in renewables.

Over our ten-day expedition, we explored how renewables could survive in the harshest climate on the planet. Upon arrival back in Ushuaia, Argentina, after two days of navigating fifteen-foot swells in the Drake Passage, I left him with a mission—become the storyteller that the whole team saw he had the potential to be. Moreover, I challenged him to inspire others through his stories to keep pushing to advance the deployment of renewables.

Eleven years later, Tom has come through with his commitment to tell the story of those original risk-takers, who like me, saw the promise of making a positive impact and committed to it. Despite the personal struggles, hardships, and constant questioning of sanity by many of those early pioneers in the industry, Tom's compilation of stories in this work goes to the heart of why renewables entrepreneurs and risk-takers are unique unto themselves.

His gifted ability to be a storyteller takes us on a journey to personally get to know the background stories that empowered these individuals to take their profession to the next level, shedding ego of oneself and surrendering to a united cause to leave this world better than how we came into it.

Our journeys in taking a risk were not linear by any stretch of the imagination. In my initial years exploring Antarctica, I was dealt a shuddering blow minutes after my team reached the South Pole. It was 1986 and The Southern Quest—the ship we had planned to leave Antarctica on—had been crushed between ice floes and sunk. Most of the team was evacuated over the following days, but two members stayed behind to look after the base camp.

The expedition, the ship's sinking, and evacuation return for my teammates left me heavily in debt. But I was determined to clear up all the equipment and rubbish left behind during the evacuation, whatever the cost. I'd made a promise to leave Antarctica as I found it.

It took a long time, and I was bankrupt, but it was worth it. It shaped my life and I've never looked back. Ultimately, despite this and many other setbacks, I was dedicated to continuing to take the risk and seeing the mission through.

Seeing the mission through requires core traits that Tom theorizes are essential to a renewable energy risk-taker. These traits range from having a sense of inspiration, dedication to cause, servant-leadership and social entrepreneurship. All of these provide a window into what it takes to take on a calling to work in the renewable energy industry.

In much the same way, from my first treks to Antarctica, I realized you couldn't approach an expedition with an ego, thinking, *I'll take this one on* because that attitude would literally kill you. Sensitivity, care, respect, humility—even love for each other—got our teams through the roughest times in Antarctica.

Tom reaffirms this lesson though his work here. In addition, he urges us to have another daily reminder "on repeat" when approaching our daily professions. Taking calculated risks is necessary for us to reach our own potential as well as the outcomes we yearn to achieve. Indeed, risks drive our ambitions, ambitions spur our dreams, and dreams in turn forge our futures.

Twenty-two years ago, Jacques Cousteau advised me to, "Focus on one thing and you will deliver something—don't go too much out." I hope I've learned that lesson by now. I ask you that when you finish reading this book, focus and never forget that if you have the ability to think it—or dream it—you have the ability to begin enacting it right now, today.

Remember to remain positive and that you have the power to overcome obstacles and achieve your goals—whatever those may be—through perseverance, determination, and commitment.

For boldness has genius, power, and magic in it.

Good luck to you all and I encourage you to join Tom in taking the risk in advancing to the next phase of the renewable energy industry's evolution.

Sir Robert Swan, OBE

CHAPTER 1

The Proverbial Oak Tree

Rana Adib of REN21 recently reflected on hydrogen and its role as a new burgeoning technology in the renewable energy sector. She said, "Don't think about the roof and then stop building the house's foundation. There is no clean hydrogen without renewable power. We need renewable power to produce clean hydrogen."

Reading this quote got me in turn to reflect on eighteen years of my own journey in renewables. I had realized I had not stopped in all those years to look back on my own career, let alone celebrate how far we, as an industry, have come.

Just as Rana noted the need not only to focus on the growth of hydrogen, but to do so understanding the importance of it being derived from renewables, I too needed to get back to my roots and juxtapose my beginnings against where I am today in my career.

I asked myself a variety of questions. Have I taken the time to pause on my current career progression to reflect on which traits got me where I am today? Have I learned from my previous professional mistakes? Have I continued to challenge myself? What risks have I taken to impact this industry?

The adage of "Don't forget where you come from" rang true in my mind when thinking back to how I started in renewables. It was time to consciously decide to carve out time to reflect on my beginnings in renewables. More importantly, I wanted to think back to those mentors who made an impact—and continue to make an impact—in me continuing my career in renewables.

With the recent passing of a mentor of mine, what first started out as an op-ed that I wanted to place in an industry magazine, turned into a much larger project. Initial conversations to pull together the op-ed led to interviews with some of the greatest risk-takers in the US industry.

I realized through these interviews that something sets apart renewables "entrepreneurs" and "change-makers," which can be ascribed to a risk-taking mind-set unique alone to our industry. Indeed, this mind-set intricately weaves risk-taking with an intrinsic value of "wanting to do the right thing" and further combines with an astute utilization of brilliant business tactics.

This unique mind-set is directly tied to a "one-of-a-kind" journey the US renewables energy industry has taken, surpassing limited initial expectations for survival back in the 1970s. Many once saw the industry's "modern" birth as a short-term "pet project" tied to the infamous oil crisis that defined President Carter's administration.

The 1980s and 1990s had others question whether the interest in solar was short-lived, only to be a blip on the energy industry's radar. As oil and gas prices settled, American geopolitics focused on ever more reliable relationships that provided access to pipelines in the Middle East. Renewables were cited as "too expensive," "too unreliable," and tied to a Democrat-leaning DC "political agenda."

The 2000s saw an ever-more-serious climate change conversation. Serious investment from Wall Street coupled with a plethora of start-ups gave the industry a much-needed infusion of capital and technologies. The 2010s saw an ever-stronger workforce with a passion for the environment, which has now given the industry a runway from which we can define our next evolutionary chapter.

One thing the US renewable energy industry got right in its years of infancy was the need to take risks in order to succeed as part of the journey. The renewables industry is notorious for a series of consistent booms and busts, opportunities and challenges, and hypes and crashes every decade since the 1970s.

Despite all this, renewables leaders kept pushing when others wanted to minimize risk. So, what do these renewables change-makers have that makes them so different? In speaking with over 125 of them, I realized three factors carved them out as unique.

The first was that their success came from calculating that the risks they were taking were acceptable ones. Many of these entrepreneurs have higher than normal pain thresholds as it comes to accepting decisions that come with higher-than-normal risk. However, they still have boundaries associated with what they considered risky.

In their minds, taking a risk that would undermine their altruistic intentions in "doing good" or "saving the planet" or that could impact the key values or beliefs upon which their entrepreneurial efforts are built, would be deemed an unacceptable risk.

Second, in their initial journeys into renewables, a "moment of truth" opened their eyes to an opportunity and an interruption in their linear journey, both of which helped define their ultimate trajectory.

Third, they chose to be in a "coalition of the willing," believing in the potential the renewables industry had, gathering with like-minded individuals, and building networks that still work together today, decades after they were formed.

All three of these traits were tied to one common theme of persistence, which despite greater public acceptance and business profitability, is needed now more than ever before.

To those of us who grew up in the 1980s, the phrase "meeting at the oak tree" connoted not only meeting at a physical location, but a philosophical common point or logic that gathered like-minded individuals together. Because of this, I wanted to create a proverbial "oak tree" or "meeting place" through this book.

My hope is that the industry in reading about these risk-takers can gather around as renewables professionals, to converse, reflect, digest, laud, and amplify those traits espoused by the leaders you'll read about. Indeed, I believe all of us should embody these key traits.

When I came into the renewables industry as a twenty-four-year-old, just as many of you are coming into it now, we had to build up our renewables market knowledge and immerse ourselves in networks from day one.

Just like any fast-growing industry, our industry was starving for talent and skill sets. Some of us started in Washington DC at trade associations and nonprofits focused on convening key stakeholders. Others went straight to Wall Street, joining investment firms and banks poised to expand market access to capital. Still others went to Silicon Valley to tackle some of the greatest technology paradigms of our lifetimes.

Regardless of our career starting points, we quickly learned over the course of our nonlinear journeys that the networks we fostered were just as invaluable as our specific technology, finance, or policy know-how. Those networks formed during our formative years in the industry have contained mentors, personal heroes, and leaders whose traits have inspired us to challenge the boundaries of our abilities and capabilities.

For these reasons, I've written this book, wishing I had many of the "nuggets of wisdom" as someone starting in the renewables industry in the initial days of my career. These nuggets may have been one-time short-term beneficiary lessons, but they have manifested into traits and skills that continue to benefit me throughout my career.

And for those of you who've been in this journey for a while, now is the time to pause and reflect, benchmark, and reassess where you are in your careers. Indeed, now is the time to ask, "In making a name for myself in the industry, have I truly challenged myself and taken risks to better the industry?"

Just as important is the question, "What impact have I made to contribute to the growth of the industry? Have I pushed the boundaries of innovation and creativity in my profession?" Last, "How have we as an industry worked to break down the barriers that hamper leadership in our industry? And how do we ensure that risk-taking is encouraged in our industry?"

The last group of professionals this book is written for are those potential change-makers looking to make the jump to entrepreneurship. These entrepreneurs are confronted with a career that contains equal moments of adrenaline-rushed excitement tinged with a sobering and humbling lack of guaranteed success.

These change-makers are focused on creating next-gen technologies—ranging from AI demand response and long-duration "super" batteries to recycling solar processes—and I hope they become inspired to make the jump into the unknown. This industry was founded, and continues to thrive, due to the constant inflow of change-makers and entrepreneurs who challenge us in the industry to keep up. We need you.

Indeed, our industry relies on our unique set of risk-takers. Our industry holds an open invitation to them as we forge our own belief in the ever-evolving potential our technologies bring to the energy market. We need to ensure that our industry retains an "open for business" mind-set, focused on invoking all those with diverse skill sets that will aid our industry's evolution this century.

As we embark upon this literary journey together and gather around the "oak tree," we'll work to better understand the unique mind-set that has driven risk-takers in renewables to persist in both challenging the status quo of our industry as well as working to make our world a better place.

Oak trees are majestic and live longer than humans do; they have a life expectancy of one hundred and fifty to three hundred years with some as old as four hundred years. That's what we want for renewables—not just to grow and profit quickly, but to endure, mature, and remain as a sector. In a word, we want to persist.

You never know. You may realize you already have the traits and unique mind-set to become part of the risk-takers needed for the next phase of the renewables story.

All it takes is for us to gather around this growing oak tree, known as the US renewables industry, and take a collective risk together. For we are an industry that is truly unstoppable due to the risks we've taken and the ones we're about to take.

PART 1

LAYING OUT
THE CASE

CHAPTER 2

Let's Start at the Beginning

We all start our careers somewhere. Sometimes that "somewhere" in no way, shape, or form relates to where your career will end up. However, in looking back, it's uncanny how things work out, with each job, position, and industry—like building blocks—providing a foundation to build your own personal career story.

Like a tailored tutorial for your career journey, which you unknowingly embarked upon, you pick up leadership traits along the way. These traits come through honing skill sets you've developed over your career as well as traits you've adopted from mentors. Plain gut instinct also plays a role.

As a young twenty-three-year-old who had just been laid off from a Silicon Valley job due to corporate restructuring, I had arrived back in Washington, DC, naive but eager to be back. Having graduated a few years earlier from Georgetown University, returning to DC was coming back home. The streets were all the same, with familiar bars—like Rhinos in Georgetown and Brass Monkey in Adams Morgan—packed with Hill staffers debating politics.

After undergraduate studies in diplomacy at Georgetown followed by graduate studies in international development at the University of Chicago, I had originally thought I'd be a diplomat. However, after taking the Foreign Service Exam four times to no avail, I realized I was approaching my future career in the wrong way. I was looking at a career end goal but was not dedicating time to creating a game-plan for how to get there.

While at the University of Chicago, and at the urging of a mentor, I homed in on my skill sets and what I enjoyed doing. From telling a story in front of a captive audience to marketing student volunteer efforts and events, I discovered by strongest skill set through communicating a story. At that point I realized marketing was my passion, and I looked forward to applying that skill set to an industry that was making a positive impact in the world.

Settling back into DC included avoiding getting pebbles in my sneakers on daily morning runs on the Mall as well as remembering that the metro always seems to have limited service. Being back reminded me how truly small DC was. This was proven by consistently bumping into someone I knew—literally on every block—on my way downtown.

"Welcome to ACORE!" were the first words I heard that late November day when I walked into a townhouse located south of Dupont Circle. Having left Silicon Valley a little deflated and feeling like a failure, those words—delivered with such sincerity and hope—defined a moment I'll never forget. I would associate those first words with my entry into renewables. The deliverer of the welcome was none other than Jodie Roussell, a former classmate and friend from Georgetown.

ACORE, which stands for the American Council on Renewable Energy, is the nonprofit Jodie had helped set up that previous year. Her own journey into renewables sprouted from nontraditional roots, given her degree in Chinese and Linguistics Studies from Georgetown.

Jodie had always been an excellent shepherd for me and foresaw something in me that I did not know of myself at that point. She knew we were the people who would eventually lay down the foundation of the oak tree known as "ACORE." Over the next years, I learned, along with Jodie, that we might just be able to do it. She joined with the pioneering start-up to plant the oak tree because she believed in the potential of the renewables industry and the ability to build networks. She took the risk.

What always struck me about Jodie was her ability to be adaptable to whatever situation she found herself in. Her perception was always spot on, matched by her warm smile and sincere interest in knowing what exciting things you were working on.

Jodie proceeded to ask, "How was California?"

Three weeks prior in September of 2004, I was packing up four boxes of all my belongings and shipping them home to Orange County, New York, from Mountain View, California. My layoff in Silicon Valley had cemented the fact that my "California Dream" had ended. Being laid off at twenty-three came as a big shock. Knowing I always would have the support of my folks, I opted to leave California and come home to Goshen, New York.

Two days into my return to the East Coast, I was sitting at my parents' kitchen island with a paper and pen going through a list of contacts I had from previous internships and college connections.

Times were different back then when jobs were found through direct in-person emails and networking versus today's LinkedIn. After laying out my job search strategy, the realization was clear. My strongest network was back in Washington, DC, and I needed to go there.

A few days later, I packed up a bookbag and duffel, gave myself three weeks, and said in a determined way to myself that I'd come back with at least one job offer in hand. Having reconnected with friends to find a place to stay during my search, Jodie had offered me her couch in Capitol Hill with the invitation to stay for a few weeks.

Before I knew it, I had stepped off the Northeast Regional Amtrak train at Union Station and was finding myself back in DC, headed to the address in Dupont Circle that Jodie had given me.

"Let's put your things on the side and let me show you around," said Jodie as she hugged me. I still remember the feeling of warmth the minute I entered the room on the second floor of the townhouse. That warmth was accompanied by the soft smell of balsam fir that hung in the air.

Jodie in her typical way read my mind. "Aren't the garlands and wreaths great? I had them shipped from my folks' place in Vermont to give the office a Christmas feel." Little did I know that fresh garland and wreaths from Vermont would be a tradition for many of my years at ACORE.

She continued, "I'm excited to have you back in Washington. I have to work for a few more hours this afternoon, but why don't we get you set up with a workstation, and you can get some emails and calls in before we head home."

I nodded, and Jodie led me into an adjacent small room filled with three to four interns working away and proclaimed, "Let me introduce you to Thayer Thomlinson, Kristin Deason and Andrew Deason."

The Deason siblings were a prime example of those entering the industry back then, sought after for their skill sets developed in other industries. Kristin was originally a defense contractor and had started graduate studies in energy and environmental management, wanting to execute a career change into renewables.

She found out about ACORE and applied for an internship to get a foot into the industry. When she interviewed with ACORE's founder, Michael Eckhart, he saw her bachelor's degree in computer science on her résumé and mentioned that ACORE needed IT support.

"I'd be happy to advise in that area a bit, but I really want my internship to focus on renewables because that's where I want to gain experience," Kristin responded, having already had her fill of IT work from her previous position.

Mike understood and joked under his breath that he probably needed someone under the age of eighteen anyway, as they'd be more familiar with new technology. Kristin retorted half-jokingly with, "Maybe you should interview my brother," who at the time was a computer whiz student at a nearby science and tech high school.

Long story short, Mike did indeed interview Andrew, and in the end, Kristen's little brother—eleven years her junior—was hired for a paid job with ACORE while she was hired for an unpaid internship. Regardless of the outcome, both siblings savored the time at ACORE, absorbing as much information as they could about the burgeoning renewables sector.

After Jodie made formal introductions, and I had put down my things, she beckoned over to me to introduce me to another key member of the team.

After passing back through the main foyer where Jodie's desk was, which had a capped fireplace and large evergreen wreath behind it, we walked into another room. This room was well-appointed and anchored by a serious looking gentleman in his forties reviewing a document.

We ushered into the room with Jodie making our presence well-known. "Roger, let me introduce you to my friend, Tom Weirich." Roger Ballentine sized me up and down in one look, smiled and with a firm handshake, welcomed me back to DC.

"ACORE is subletting space from Roger, who started a consultancy after a number of years at the White House under the Clinton Administration," Jodie explained. He is also on ACORE's board of directors and has been helping Mike and me start the organization up."

I'd find out and experience over the coming months that Roger embodied the trait of servant-leadership, having dedicated his career to working with the US government to enact changes to clean our water and air. In addition, he provided mentorship to many of our current civil servants that continue to lead his charge.

Roger had an impact-filled previous career as a senior member of the White House staff, serving President Bill Clinton as chairman of the White House Climate Change Task Force and as Deputy Assistant to the president for Environmental Initiatives. Prior to being named Deputy Assistant, Roger was Special Assistant to the president for Legislative Affairs, where he focused on energy and environmental issues.

After his government career, his commitment to being a servant-leader manifested itself in numerous ways. Over the years, Roger had helped some of the world's leading companies and institutions increase competitiveness and manage market and regulatory risk through cutting-edge energy and sustainability strategies.

Using his expertise and deep relationships in government to navigate through what, at times, were challenging political situations, Roger had helped many up-and-coming start-ups, as well as big-name corporations. He worked with them to develop better business strategies, make better investment decisions, and negotiate new business partnerships.

He put the renewables industry first and served as a connector, fostering greater public-private partnerships by building critical alliances, diverse energy stakeholders, and then developing and implementing renewable energy and other sustainability measures that would achieve corporate buy-in and yield bottom-line value. In many ways, we have early policy pioneers like Roger to thank for the current Fortune 500 corporate procurement push.

With this introduction, I had gained one of my first mentors in the industry. Having exchanged the necessary formalities and discussed the job search reason why I was bunking at ACORE for two weeks, I headed to Jodie's apartment to begin plotting my next career steps.

Finding a job in 2004 was different from today's approach to finding a job via LinkedIn. It was all about who you knew, meeting in person or making a phone call, demonstrating your eagerness and hunger, and reconnecting with influential contacts who could vouch for your work ethic. In many ways, it was more direct and easier, but I was relentlessly persistent—a trait that would continuously come in handy when working in renewables.

I took out my laptop, which weighed like a brick, and began to call contacts to let them know I was back in DC. Before the age of smart phones, texting, and a more advanced LinkedIn, this was the most expeditious way to get a hold of your rolodex.

Reaching out to folks in my network, I connected with a few public relations and advisory firms and found myself within a week in four separate interview rounds. As my two weeks in DC were rounding out, I was ecstatic, having ended up with three pending job offers.

Starting to pack up my duffle at Jodie's house, I was looking forward to delivering the news at home as well as doing a much-needed load of laundry.

Jodie walked into the living room when I was zipping up my suitcase with a contemplative face. After I asked her what was wrong, she confided that ACORE had an upcoming policy forum on Capitol Hill at the Capitol Hill Club, and she didn't know how they would manage staffing the event.

She said she had her three interns to assist, and Mike had his two daughters sign up to manage the registration desk. Despite the help, she didn't know how to handle the media and the sponsors onsite, which included initial renewables heavy hitters GE Energy Financial Services and enXco.

A Harvard MBA graduate, Mike came from a distinguished energy and investments background into a world where renewable energy deals simply did not flow, despite immense potential. Seeing that opportunity would mean an interruption in his own linear career journey, and he'd come to Washington, DC, to do something about it.

He gathered up pioneers and found all the risk-takers, one by one, knowing they would become the oak tree known as ACORE—an organization that would forever change the face of renewable energy finance and policy. Mike branched out to integrate all parts of the renewables industry and believed the real impact would be felt if the industry was heard on Capitol Hill.

Coming back to the task in front of us, Jodie and I sat down and walked through the event. After a few minutes, I volunteered to help.

"Heck, you let me sleep on your couch for two weeks. It's the least I can do," I remember telling her. Little did I know that changing my return plans to New York that night was yet another step in my journey into the renewables industry.

The next morning, Jodie walked me over to International Square, a large office complex close to Foggy Bottom, to introduce me to Mike. Walking into a Regent office suite, which was a shared office workspace that was a predecessor to today's "WeWork," we walked down a few twisting hallways. With a knock, we entered an office filled with some bookcases, and one heavy-looking wooden desk in front of us.

Mike stood and welcomed us, shaking my hand. Jodie excused herself, stating she was running late for some intern programming and had to get back to the townhouse. Mike and I then sat down for a conversation.

He asked me about my background and how I knew Jodie followed by a conversation on ACORE's Policy Forum in detail. Walking through the event, Mike coached me on its significance for being the first time the entire renewables industry would be convening on the Hill as well as the various policies being spoken about.

The ACORE Policy Forum was held in conjunction with the Environmental and Energy Study Institute (EESI) and the Renewable Energy and Energy Efficiency Caucuses of the US Senate and US House of Representatives. The event kicked off with an Opening Reception on December 6, 2004, sponsored by General Electric (GE). Jodie, the interns, and I were at the Capitol Hill Club by 6 a.m., setting up for that evening's reception and for the next day's forum.

Powered by more than a few cups of coffee, we laid out name badges, put up signage, and made sure the IT team at the Club had set up the projector and screens correctly. One by one, countless conference attendees arrived, each with more impressive titles than the prior. After the majority of participants was upstairs, Jodie told me to head up to make sure sponsors were entertained and attended to.

After a few hours of networking and assisting with sponsor items, I finally took a seat to absorb it all. With acronyms like the RPS, ITC, PTC, FERC, and RECs being touted left and right, I sat back trying to piece together the conversations. I digested important drivers for the early industry and, looking around the room, memorized faces and names for potential conversations regarding job openings.

Sensing my bewilderment and attempt to absorb everything, a distinguished-looking gentleman sat down next to me and, with a handshake, introduced himself as John Mullen. John, alongside Mark Riedy, were ACORE's initial attorneys, working with Mike and Jodie on the nonprofit's 501(c)(3)'s filing. In addition, they served as trusted mentors and de-facto advisors to ACORE's team.

"So, what do you think about all of this," asked John with a smile. Still digesting and overwhelmed with everything around me, all I could answer was, "It's a lot to process, but I can sense the passion everyone is bringing to their presentations and messaging to the government."

In the first day alone, curiosity and a sense of adventure had encouraged me to get to know this budding industry further as I could see myself in it.

John keenly sensed my interest and took me under his wing that evening, introducing me to many of whom we consider "the legends of our industry" standing around the Capital Hill Club.

At that moment, John had become yet another of my first mentors in renewables. He showed me that all it takes is setting up one or two introductions to ignite a cascade of conversations that can impact a young person's professional career.

Despite our opposing political views outside of renewables today, I have continued to cherish my relationship with him, filled with respect and constant openness to a hearty dialogue or two every time I see him and his wife Gwen in Alexandria, Virginia.

The following day, the ACORE Policy Forum was a success, with over 150 industry participants that included key Hill Congressional and Senate officials, their energy aides, and numerous journalists from mainstream news outlets. After a long day, it was time to head up to the third floor of the Club for a private closing reception.

I sat down along the side of the room to take it all in. I'd heard inspirational speeches about investment in renewables and the need to have investors and the financial community take renewables seriously. The conversations around me also centered on the need for the renewables industry to stop being fragmented, with the solar, wind, geothermal and hydropower sectors going against each other. Leaders who spoke earlier that day at the conference were advocating for a united front.

Many were also calling for the federal government to take renewables seriously and structure policies—such as investment and production tax credits as well as renewables portfolio standards—that would provide a firm foundation upon which to build a twenty-first century industry.

Indeed, panel session after panel session filled with speakers who defined the industry had awakened a curiosity about this "renewables industry" I knew very little about.

Only months later did I appreciate the range of speakers who were the "the firsts" of the industry. Present on site were leaders such as Amory Lovins, Founder and CEO of the Rocky Mountain Institute (RMI) as well as Robert (Bud) McFarlane, the former National Security Advisor to President Reagan.

Others invested in the dialogue included Dan Reicher, who had come out of the Clinton Administration as Assistant Secretary of Energy for Energy Efficiency and Renewable Energy as well as having served as the Department of Energy's Chief of Staff. Steve Zwolinski, President GE Wind Energy was another notable figure onsite.

These leaders, along with more than twenty other industry headliners, focused on solutions—versus barriers—coupled with requests for policy-makers, financiers, and technology providers to unite and forge a united renewables' destiny in the energy sector. j

This chance encounter with renewables through the ACORE Policy Forum had me thinking. I was naturally a bit uncomfortable venturing into a new industry at first, but I was about to find out that putting myself out there and lunging into renewables would reap many rewards.

With my head still spinning after two packed days and a good night's rest, I found myself in Mike's office again. This time, I was committed to deliver a direct pitch to him. "Mike, I'd like to work in renewables and join you and Jodie at ACORE."

Mike sat back in his chair behind his desk and paused. "We don't have a way to pay you, but I see you're sincere and a hard worker, so let me speak with the board and get back to you." Little did I know that my persistence would pay off. The next day, Mike called me back with great news.

"We have a partnership for a new trade show—the first all-renewables exhibition and conference, called PowerGen Renewable Energy (PGRE). The event is run in collaboration with PennWell Corporation. PennWell has someone on exhibit sales, but we're not getting the sales figures we want. We can offer you a short-term consulting job selling exhibit booths and marketing the event. Would you be interested? PennWell would pay us to pay you, and you'd be on a small base salary with a high potential bonus payout if you do well."

After some back and forth, I accepted. With that, and unbeknownst to me in the grand scheme of things, I was beginning my first day in what would be my lifelong career in renewables. I had finally matched my skill set with an industry I saw had immense potential, and I wanted to be a part of it.

CHAPTER 3

The Rollercoaster

In collaboration with Michael Liebreich

One of my favorite articles of all time was authored by industry expert Paula Mints. It appeared in *Renewable Energy World*, comparing the renewable energy industry to a roller coaster ride back in 2016. Citing the parody, she wrote:

Before boarding theme park roller coasters, riders typically see the following signs:
- For your safety remain seated with hands, arms, feet, and legs inside the vehicle.
- This is a high-speed roller coaster ride that includes sudden and dramatic acceleration, climbing, tilting, dropping and backward motion.
- Beware, you may lose your glasses and hats on this ride.
- You must be this tall to ride.

Similar signs are neon-light visible in the renewable energy industry. Essentially:
- Proceed carefully and thoughtfully and keep your emotions in check when making business decisions.
- This is a high-speed volatile, incentive-driven industry with sudden market changes due to withdrawal

of government funding and painful downward price pressure.

- Beware, you may lose a ton of money on this ride.
- You must clearly understand and accept the difficulties of competing in an incentive-driven industry with severe downward price pressure for components and downward bidding pressure on tenders.

Paula ended her piece with, "It is not always true that the best business plan and the best technology will win, but it is true that the [renewable energy] industry and all of its technologies and participants [are] crucial to the future."

***

Michael Liebreich and I first met at Renewable Energy Finance Forum (REFF—Wall Street) back in 2005 when I started out at ACORE, and he was ramping up New Energy Finance (NEF). Over the following months, we worked together on a joint industry newsletter to serve the US market, which was a precursor to BNEF research briefs.

I also had the daily pleasure and fortune of having Ethan Zindler, his first hire here in the US, down the hall from where I sat, when NEF shared space with ACORE in those early days. Michael had asked Jodie and me to assist him in interviewing candidates for his first North American hire.

From the moment Ethan walked through the door, we were impressed. Ethan's experience included stints at the White House, MTV, and as a freelance journalist covering two World Cup soccer tournaments. Though he had no renewables background, he had a zeal and passion for journalism, data, and reporting, and he believed in the mission of NEF and the broader renewable energy industry.

Recollecting his early pioneering days, Michael's approaches to calculating risk and "taking the chance" were tied to the critical events that shaped the renewables markets globally. We had enjoyed a conversation juxtaposing our journeys against the backdrop of the US renewable energy industry's experience since the 1970s and agreed that this background tested the original risk-takers in the industry.

In mapping out those events, and how they offered opportunities for our own innovation and entrepreneurship, Michael shared his thoughts with me to incorporate them into the thesis of this book.

To start, we need to paint the backdrop surrounding the narrative around energy transition, which is nothing new for the US energy industry. The conversation surrounding the need for a greater energy revolution started as far back as the turn of the last century. William Moomaw, lead author of a chapter in the IPCC report on energy options and professor of internal environmental policy at Tufts University, sat down with *The New York Times* in May of 2007 to reflect on the need for an energy transition in this century.

In his interview, Moomaw believed we were on the cusp of a significant milestone in energy:

> *Here in the early years of the twenty-first century, we're looking for an energy revolution that's as comprehensive as the one that occurred at the beginning of the twentieth century when we went from gaslight and horse-drawn carriages to light bulbs and automobiles. In 1905, only 3 percent of homes had electricity. Right*

now, 3 percent is the amount of renewable
energy we have today. None of us can predict
the future any more than we could in 1905, but
that suggests to me it may not be impossible to
make that kind of revolution.

Indeed, none of us can predict the future given that we're now approaching 22 percent renewable penetration in the US, according to the US Energy Information Agency. However, we can learn from the systematic drum beat of the markets that power the overall energy sector.

The US renewable energy industry has, for as long as many of us remember, gone through a "boom and bust" cycle of the "modern" renewable energy industry, which to many began with the Oil Crisis of the late 1970s under the Carter Administration. And the cycle continues even today.

This cycle can be defined by periods where policies and incentives, financing structures, pricing of technologies, and public acceptance of renewables fluctuated, impacting the penetration and scalability of the US industry. Many of the executives, leaders, technologies, and project structures that I've researched for this book began in the late seventies and, despite the cycle, persisted in their endeavors.

In order to set the stage to walk through the boom-and-bust cycle from the 1970s to the present, one needs to take a big-picture view and approach this industry's global history through five broad phases. In working with Michael on this history, I've included additional independent research that uniquely shaped the US experience in renewables vis-à-vis the global experience.

Despite this "boom and bust" rollercoaster cycle, we saw renewables entrepreneurs and risk-takers in the US not only survive—but thrive—by taking advantage of the highs and lows that provided capital, research and development, and innovative responses to the challenges in technology deployment.

PHASE I

According to Michael, the first phase from the early 1970s to the turn of the century was focused mainly on research and development. Renewable energy was extremely expensive, and it could only be implemented commercially in very limited applications such as satellites, pocket calculators and parking meters.

In the 1970s we saw the first environmentalists mobilize. On April 22, 1970, Denis Hayes helped to inaugurate the first "Earth Day." That initial gathering raised awareness of the threats to the environment and was held in hundreds of cities and towns across the US, attracting a reported twenty million people.

The public had become more aware of the various previously unknown health and safety costs associated with traditional forms of energy and were beginning to demand more of the US government, which resulted in the creation of the EPA in December of 1970. The EPA began to issue policies that for the first time valued those costs by associating historic carbon metrics to dates by which to roll back carbon and GHG emissions.

During the years of 1973 and 1979, the United States experienced significant petroleum shortages as well as elevated prices. The roots of the oil crisis began in the 1960s according to a *Time Magazine* piece called "Oil Squeeze," when the US and most top petroleum producers, such as Germany and Venezuela, had experienced peak home production. These peaks began to put upward pressure on oil prices globally, and during this period the US became increasingly more addicted to petroleum from the Middle East region, specifically from Saudi Arabia and Iran.

The oil crisis of 1979 was precipitated by the massive protests and eventual Iranian Revolution, which completely upended the Iranian oil sector. While the new Iranian regime resumed oil exports, they were inconsistent and at a lower volume, forcing prices to go up. Saudi Arabia and other OPEC nations increased production to offset the decline, and the overall loss in production was about 4 percent.

Panic ensued in the US, which drove higher than normal prices, and many Americans, seeing our addiction, started to investigate alternative routes to powering the US using home grown energy. Americans began to explore putting solar on their roofs, as well as investigating the potential of blending ethanol in fuels to stretch reserves. President Carter, wanting to show a visible commitment to renewables had thirty-two solar panels installed on the White House West Wing roof in the summer of 1979.

This period was unfortunately short, and despite periods in the 1980s of invasions and wars surrounding Iran and Iraq, oil prices began to decline as other countries stepped in to backfill petroleum production deficits. A spark in the renewables sector, specifically in solar energy and biofuels, had been extinguished as Americans went back to their normal lives using traditional fossil-fuel-based energy.

The 1990s came with some incentives in the form of limited tax credits. However, the general population in the US still considered renewables a "luxury." With the onset of the Gulf Wars, the kernel of doubt had been planted, and many started to ask whether renewables would become a much more needed part of the country's future energy mix.

This was encapsulated famously in a 1995 *Wall Street Journal* piece by expert energy journalist Agis Salpukas entitled, "70s Dreams, 90s Realities; Renewable Energy: A Luxury Now. A Necessity Later?"

In the article, Salpukas described how "the so-called renewable sources may well survive only as window dressing for utility company annual reports" because of price competition. "Renewables are a luxury that belt-tightening utilities can no longer afford, executives complain, particularly now that natural gas is so plentiful and gas-fired generating plants so inexpensive to operate," he went on to write.

Salpukas further noted however, that "the decline of renewables carries long-term risk, for example, if global warming becomes something more than just a threat."

Another key historic milestone that marked the 1990s was the creation of a corporate entity that gave birth to the renewables careers of many of the leaders in the sector—Enron.

Enron, an energy company that was based in Houston, grew in tandem with the deregulated market of the nineties. Over the years, as a result of accounting loopholes, it set up special purpose entities and, with intentionally "poor" financial reporting, was able to hide billions of dollars in debt from failed deals and projects. This ultimately led to its demise and bankruptcy in 2001.

PHASE II

The second phase came into play in the late 1990s and into the early 2000s, with Germany's historic Erneuerbare-Energien-Gesetz (EEG, Renewable Energy Sources Act), driven through by the force of nature that was Bundestag Member, Hermann Scheer.

Taking a cue from Germany, during this time we saw the renewable energy industry amass here in the United States through the creations of energy corporations and the launching of renewables IPOs.

This period was also known for two events that forever altered the trajectory of the industry and the career of many of the leaders in our industry. The first being the deregulation of the market and the ensuing California Energy Crisis. The other being the rise and fall of infamous energy giant, Enron.

The years of the late nineties in the move from Phase I to Phase II saw the overall energy sector move to become deregulated and the wholesale electricity market headed in that direction also. In this instance, the phrase "wholesale electricity market" related to the purchasing and selling of large quantities of electricity between utility companies.

Although wholesale electricity could not be implemented by renewable energy sources in many states—specifically California—it was taken advantage of by newly formed energy giants and led to a crisis that left scars on the sector for years to come.

During the early 2000s, Enron's hunger to grow into an energy giant led it to be a hub of innovation, attracting talent and entrepreneurship around what was a nascent renewables industry. It took on the technological challenges associated with solar, wind, and burgeoning renewables. Enron furthermore centralized financing, development, manufacturing, distribution, and installation of these technologies under one roof.

Through Enron's investment in the US renewables sector, it helped bolster the value proposition of renewables to the energy market, expedite R&D around next generation technologies, and create supplier and developer networks that gave birth to additional renewable energy "giants" that still exist today.

Despite all this growth, the California energy crisis of 2000–2001 rocked the energy sector. California had found itself having a shortage of electricity due to market manipulations and capped retail electricity prices. The state suffered from multiple large-scale blackouts. Utility giant Pacific Gas and Electric Company (PG&E) went bankrupt and Southern California Edison was near bankruptcy.

The causes of the crisis were tied to several factors according to the US Energy Information Administration (EIA). Drought played a factor along with approval delays of new power plants. And remember Enron? Well, they played a factor in a market manipulation that decreased supply, causing an 800 percent increase in wholesale prices from April to December 2000. Despite California having an installed generating capacity of forty-five gigawatts, demand was around twenty-eight gigawatts at the time of the blackouts.

In this instance, we're measuring output of energy by GW. A gigawatt, abbreviated as GW, is a unit of electrical power equal to one billion watts (W). These units are most commonly used in referring to the total output of a type of energy production like solar energy or wind energy but can also be used to measure the total energy capacity of a state or country.

A gigawatt is a massive amount of energy equaling a thousand megawatts. To put this in better perspective, that is enough energy to power a medium-sized city. According to residential solar installer Sunrun, to generate that much power in a day would take over three million solar panels.

The resulting demand-supply gap, created by entities including Enron, was facilitated by energy traders who took power plants offline for maintenance during peak demand to artificially create a shortage and drive up the price.

In certain instances, these practices drove electricity prices up to a factor of twenty times its normal value. This resulted in a squeeze in industry revenue margins, which were tied to the state's cap on retail electricity charges. The impending collapse of the energy market loomed.

From the decade between 2000 and 2010, renewable energy grew in scale and key capabilities were built. Growth specifically occurred in the supply chain among utilities and network operators as well as in the financial system.

However, the technologies remained expensive and required subsidies or other forms of policy support, which in the US consisted primarily of the investment tax credit (ITC) for solar power and the production tax credit (PTC) for wind energy.

Incentives were running their course and producing favorable economic growth indicators for the US industry, but they were set to expire after a few years, only to be extended weeks—and sometimes days and hours—before they were set to expire. These jerks in incentives created a boom and plateau scenario, which continues to this day as a result of uncertainty surrounding the extension of the ITC and PTC.

During this time, specifically 2003–2006, we also saw a move from just focusing on renewables' impact on the utility scale development to the creation of the distributed generation sector, which focused on onsite opportunities to generate electricity at the source of energy demand.

This move opened the door for new entrants into the market, primarily in the form of corporates and retail giants as renewables offtakers. These entities provided demand for clean, carbon neutral, "net zero" renewables across the US.

Despite all these gains, the end of this prosperous decade came to a halt with the 2008 financial crisis. Though many will claim the effects of the 2008 financial crisis on the US renewables industry were minimal, they did take a toll. During this time, we saw China outpace the US in the solar industry, as well as VC funds dry up, curbing the building of US solar panel manufacturing plants.

In addition, financing for companies charged with production of "advanced fuels" halted, putting an end to the growth of the biofuels sector that had been empowered through the Congressional national biofuels mandate of 2007.

"A lot of enthusiasm and momentum that had been built up in the venture community for solar took a substantial hit post market crash. VCs thereafter simply did not have the capital available at the scale of the factories they needed to fund," reflected venture capitalist Stephan Dolezalek in an interview with Ben Geman of Axios a decade after the crisis.

PHASE III

This new phase began on the heels of the upswing from the financial crisis, starting in 2009. During 2009, under the Obama Administration and the American Recovery and Reinvestment Act (ARRA), policies enacted by the Bush Administration began to transition into a much more active "deployment phase." Between 2009 and 2010, renewable energy consumption rose by 6 percent to over eight quadrillion Btu.

Taking a step back to understand a Btu, the British thermal unit is a measure of the heat content of fuels or energy sources according to the US Energy Information Administration (EIA). It is the quantity of heat required to raise the temperature of one pound of liquid water by one degree

Fahrenheit at the temperature that water has its greatest density (approximately thirty-nine degrees Fahrenheit).

Energy or heat content can be used to compare energy sources or fuels on an equal basis. Fuels can be converted from physical units of measure, such as weight or volume, to a common unit of measurement of the energy or heat content of each fuel. The EIA uses Btu as a unit of energy content.

To put a Btu into greater context, a single Btu is very small in terms of the amount of energy a single household or an entire country uses. For example, the US used about 92.94 quadrillion Btu of energy in 2020.

Total US energy consumption rebounded by 4 percent to nearly 98 quadrillion Btu, due in some measure to economic recovery from the 2008 financial crisis. Rounding out 2010, renewable energy provided 10 percent, or 425 billion kilowatt hours (kWh) of electricity, out of a US total of 4,120 billion kWh.

Journalist Ben Geman noted that in many ways, the "bust" of the crisis proved to be beneficial in providing the momentum needed for renewables expansion in the 2010s.

According to Geman, "The 2009 stimulus law funneled $90 billion into low-carbon energy initiatives, including grants for renewable electricity development in lieu of tax credits. Those grants proved vital. Longstanding tax credits are key to building wind and solar power projects, but the tax equity market collapsed alongside the financial sector." He added that, "The Treasury Department quickly launched the grant program, which eventually provided over $26 billion ."

Legal expert Keith Martin validated this perspective when speaking to Ben in his article, stating that, "Deployment would have slowed down dramatically because these companies would have had to seek other financing, and it would have been more expensive even if it was available."

Starting in 2010, something critically important began to change and the third phase was in full force. The main thrust of policy switched from feed-in tariffs to reverse auctions, which had the effect of revealing the massive cost reductions industry could achieve.

Within remarkably few years ranging from 2010 to 2019, Michael observed that, "renewable energy had reached the long-awaited goal of 'grid competitiveness,' with the ability to sell without subsidies at prices below liberalized wholesale markets. The financial sector, led by financial leaders like Warren Buffett, responded by turning clean energy into just another asset class, held by many investors alongside their real estate and infrastructure portfolios."

PHASE IV

Phase IV, which began in 2020, is well on its way, and I estimate will take a decade. It's driven by three megatrends.

According to BlackRock, the world's largest asset manager, the difference between a regular trend versus a megatrend is the impact they have on permanently shifting the paradigm around investment decision-making. In addition, they claim that unlike standard trends, megatrends "don't exist in isolation; they collide and overlap with new investment themes appearing."

To BlackRock, "Megatrends are structural shifts that are longer term in nature and have irreversible consequences for the world around us. The awareness of megatrends in investment processes offers real insights. Because of that, megatrends influence our investment decisions—from the businesses, industries, and countries we invest in to the way we go about finding opportunities."

The first megatrend we're observing in this phase is that newly built renewable energy will not only be cheaper than newly built coal, gas, or nuclear power, but cheaper than running existing thermal plants. Michael claims that, "This will drive the stranding of fossil fuel assets on a huge scale and result in increasingly desperate political pushback by incumbents."

Second, we will see incredibly rapid developments in those complementary technologies, based on our experiences with them in the early 2000s, which allow for higher levels of renewable energy to penetrate the power system—demand response, power storage, long-distance interconnections, and pervasive digitization.

And third, the electrification of other industries will pick up speed—most notably in transportation—starting with cars, buses, and delivery vans but extending throughout all short and medium-distance vehicles but also in the heating and manufacturing sectors.

We must not forget that electricity currently accounts for only 20 percent of final energy use, up from 10 percent in 1970 according to the International Energy Agency (IEA). Indeed, the biggest decarbonization challenges lie outside the current remit of the power system and depend on the evolution of the power grid. Risk-takers in the form of technology pioneers advancing storage, demand response systems, and grid modernization will step in here.

This fourth phase of fossil fuel stranding will drive the penetration of renewable energy toward 70 percent or 80 percent of global power generation, according to Michael, irrespective of policy support. Of course, plenty of uncertainties remain regarding the trajectory in different industries and regions but—absenting a black swan like economically viable small nuclear reactors or fusion power—the direction of travel is already 100 percent clear.

PHASE V

A potential fifth and final phase will then be the "Mopping-up period, where the final 20 percent or 30 percent of fossil power is removed from the system or its emissions captured, and other hard-to decarbonize sectors are dealt with," states Michael.

The key question that remains now is how to accelerate the process. Countries will largely achieve their existing commitments under the Paris Agreement, but these are not adequate to hit a 2C carbon budget, let alone 1.5C, according to conversations at the United Nations Framework Convention on Climate Change (UNFCCC).

Discussions of how to achieve 100 percent decarbonization—the Phase V challenge—do nothing but distract. The Phase IV challenge is to get renewable energy rolling out at such scale that it drives fossil fuel off the grid, not merely supplementing it, around the world.

The single biggest levers to accelerate the shift to renewable energy will be campaign finance reform and the international fight against corruption. The economics of renewable energy technologies are now so compelling that they can stand on their feet but not against a deck stacked against them by bought-and-paid-for politicians. Fossil fuel subsidies have to go.

According to the Brookings Institution, globally, governments spend more than five hundred billion dollars on subsidies for fossil fuels that contribute to inefficiency, inequity, and negative externalities. Externalities—including air and water pollution, not just the social cost of carbon emissions—have to be priced into decisions.

Carbon fee and rebate systems—as championed by Ted Halstead and being tested by Justin Trudeau's administration in Canada—looks like a promising initiative. The tax-based underpricing of shipping and air travel has to stop. The mega-project-based development model that has reigned supreme since the 1970s has to be retired.

The financial sector has a huge role to play. "If we do not take action, and if emissions continue on their current path, then by 2030 we will have used up the entire remaining carbon budget. What this means is that by 2030 we will either be in a world with rapidly dropping emissions—and the only good investments will be in companies leading that trend—or we will be in a world where any greenhouse-gas-emitting business has no social license to operate," Michael notes. Investors have only eleven years to restructure their portfolios to prepare for this world—or "face the wrath of a million Greta Thunbergs."

The October 2018 IPCC report on 1.5C made clear that for a 1.5C trajectory, global emissions need to shrink by 45 percent by 2030. Even for 2C, emissions must be reduced by around 20 percent by 2030.

"So, from today, the time for funding green projects alongside brown ones is over," declares Michael. "From today, only investments consistent with a climate-compatible emissions trajectory are acceptable from a risk perspective. From today, it is no longer a question of mainstream finance versus green finance: it is a question of mainstream finance versus brown, non-risk-managed finance."

In this last phase, it is clear we need more risk-takers to get us going. We must foster entrepreneurship and leadership on climate action globally. We must continually embrace our risk-taking origins by calculating the risks, seeking opportunities to expand the market, and persisting.

The journey will not be linear, and interruptions from the financial sector are inevitable. We need to believe in the potential of the renewables industry and continue to gather around the oak tree, building networks that will take us through to the next evolutionary phase of the industry.

CHAPTER 4

Renewables Are Not for Everyone

The renewables industry is known to be very welcoming. Its eagerness to recruit and assimilate entrepreneurs and professionals that bring needed skill sets to the table has made it an amazing industry overall. In the early days, that ranged from technologists to financiers and marketers like me.

However, I'll be controversial in stating that the renewables industry is not for everyone. Indeed, professionals will enter the industry and discover it's not for them. The reason? The answer is tied to two things: (1) When you work in renewables, it's not just a job. It's a calling; and (2) Certain personality traits, when joined with this calling, form a special type of entrepreneurial leader that cannot be found in any other industry.

The industry's early beginnings were tied to two main sectors that fed it talent, specifically the traditional oil and gas energy sector and the finance sector. Both sectors culturally had a "good old boys club" mentality inherently built into them that transferred into renewables in its early days.

Another mentality that translated over was the call to tie every business model to monetization of some sort of commodity in order to spur as much profitability as possible from day one. These old mentality tenets remained with the industry throughout the nineties and early 2000s, and some will claim they are still inherently woven into the fabric of our industry.

Those making a jump into the early renewables industry had career beginnings that were unique in their own right. But one thing was common. They were all drawn to one another for some reason, with the traditional "seven degrees of separation" being cut down to one-to-one connections.

In the early days of the industry, these like-minded risk-takers came together to the oak tree to form more than a community—specifically an alliance—a "coalition of the willing"—a phrase used by many of the early risk-takers I spoke to. This alliance was built on the belief that even though all these renewables entrepreneurs had businesses to run, they were all dedicated to much more that turning profit. They were tied to a mission.

The aspirational mission and personal motivations that united them took many forms—from fighting the effects of climate change to taking on the deconstructing of traditional energy models. However, this unshakable feeling of wanting to create impact, connected with a special set of entrepreneurial traits, ultimately connected them all. Thus, a movement of individuals wanting to genuinely change the world we live in was born.

Unknowingly to them, these risk-takers were embarking upon creating an industry dedicated to turning off the old mentality of how to do business and turning on a new mentality—one that strives to create an industry that is constantly reinventing itself.

At the core of each renewable professional was the ever-torturous desire to challenge the status quo, change the planet, and convert energy impoverished societies into prosperous ones. A common trait across these risk-takers was their approach. They took that first step, mapping out a vision and keeping themselves honest, accountable, and obsessed with their final goal.

These goals have changed as the industry has matured, with conversations shifting from making "renewables part of the mainstream" to more self-reflective dialogue on how we as an industry are not only preaching our organizational values but also structuring our businesses by them.

New baselines for achievement and metrics for success have been developed by renewable energy corporations and organizations such as the Global Reporting Initiative (GRI) and the Solar Energy Industries Association (SEIA). These new guidelines judge us by the very values we espouse—ranging from GRI's Sustainability Reporting Standards related to ESG to SEIA's new Solar Supply Chain Traceability Protocol, which traces ethical sourcing in supply chains and recycling of decommissioned solar panels, storage batteries, and wind turbine blades.

We should never stop pushing the aspirational envelope of how we do business. The motivation to do better by the planet was intricately woven in the very fibers of our ethos from the start of our industry, and we cannot forget those roots.

Such goals and aspirational motivations associated with winning renewables teams can be tied back to the dynamic leaders that lead them. As much as we try to claim renewables leaders are overarchingly well-balanced, they all lean or rely on a specific definitive trait or set of traits that guarantees them success in whatever venture they take on. We are gifted with an abnormally high quantity of driven leaders and inspire us daily. So why is that?

ARCHETYPES

The concept of the modern leadership archetype has been explored extensively by anthropologist and psychoanalyst, Michael Maccoby. Maccoby says four leadership archetypes are vital to the establishment of entrepreneurial ventures.

Each of these archetypes hold a distinctive role in creating a successful enterprise. If you were to look at any profitable renewables venture, it's easy to identify each of these archetype in action at the helm. Moreover, a common laud in the renewables industry is "having a strong team," which includes a collection of executives who derive their strength from each of these archetype personalities.

Maccoby's four key archetypes can be described as the: (1) visionary leader, (2) operations-obsessive leader, (3) marketer, and (4) charismatic leader. Each of these archetypes brings a different value to leadership of a renewables organization.

The visionary leader provides vision and aspirational direction. These executives provide the "big picture" that draw employees to believe in a mission or goal. They are not controlled by their environment, rather they create their company's future.

The operations obsessive leader spends their time structuring an organization and perfecting processes that are tied to production of deliverables as well as the efficient running of an enterprise. They are the "number crunchers" and "problem solvers" who like to break up big challenges into piecemeal sizes and logically solve item by item. Board members like these types of executives especially.

The marketer knows their product and their industry well, strategically positioning a company to stand at the intersection of industry trends and customer needs in order to capitalize on the opportunity.

Last, but not least, the charismatic leader has an ability to provide vision, but at the same time, gain a followership. Gifted orators, these leaders gain affirmation of their achievements from the market and present a confident vision of what's in front of the industry.

THE PRODUCTIVE NARCISSIST

For those leaders who master their archetypes, Maccoby goes on to elevate them to the status of "productive narcissists."

Though he portrays them as narcissists who inflate their mission or their role in their own industry, he states that, "They are innovators and experts in their industries, but they go beyond it. They also pose the critical questions. They want to learn everything about everything that affects the company and its products."

While we might look at narcissists with a negative connotation, I agree with Maccoby that productive narcissists have, "Audacity to push through the massive transformations that society periodically undertakes." In an industry that has gone through so many peaks and valleys, booms and busts, without these productive narcissists in the US renewable energy sector, we would not have the industry we have today.

In reflecting on Maccoby's research, I began to ask myself whether it is possible that unlike other industries, the renewables sector here in the US has attracted an abnormally higher quantity of a specific set of these archetypes?

I claim the answer is yes. We as an industry have been gifted with both visionary leaders and charismatic leaders who have productive narcissism written all over them. The very nature of our technology and the impact it has on the world around us had attracted risk-takers to our industry who think big, think differently—i.e., outside the box—and act boldly.

Unlike charismatic leaders with nothing to back up their eloquent words, renewables risk-takers came armed with a contribution to society in the form of a mission to create a more sustainable world through the deployment of clean energy technologies.

Driven by a "high risk, high reward" mentality, renewables risk-takers have taken on the tendencies of Maccoby's productive narcissist but have neutralized the narcissist part of the equation by holding themselves to the same high standards they ask of their team.

In addition, we've been fortunate to attract risk-takers who, despite the counterculture nature of renewables, were able to challenge traditional energy norms and create the foundations for successful start-up ventures, such as Form Energy and Antora Energy, that we read about daily in Canary Media. Additional strides are also being made by the large commercial and industrial players committing to renewables including the likes of Facebook, Goodyear Tires, and Microsoft.

Looking back at my own career track, I am a marketer, one of the four leadership archetypes, and can relate well to Maccoby's productive narcissist. I jumped into renewables without the proverbial parachute to help create the ACORE brand. In so doing, I threw myself passionately into also creating a brand for renewables in the broader US energy market.

My marketing journey has come with high stakes at times but ultimately with many wins, despite my struggles with messaging about the "boom and bust" cycle of the industry. In many ways, the renewable energy industry has been a dysfunctional one with many of us describing it as "organized chaos." But it has always achieved its goals.

At times it was not good enough in renewables marketing to "just know the product and industry well." I had to have a belief that despite the finance recessions, ups and downs of tax credits, and scarcity of rare-earth resources, we would innovate ourselves out of our challenges. The one thing we as productive narcissists had going for us was the promise we now evangelize: Renewables will one day indeed become our primary energy source.

METHODOLOGY

However, to reduce renewables risk-takers' achievements to predetermined genetic archetypes or to productive narcissists' tendencies without exploring the underlying unique traits and aspirations that embolden them would be doing an immense disservice to understanding what makes them unique.

I wanted to understand this uniqueness. With every interview came a number of stories that—though entertaining to us today—reflected the struggles, challenges, and realities faced by those who embarked on the renewables journey and opened the door for all of us to follow through.

Choosing who to interview and which traits to specifically call out was not an easy task. But as with all journeys, one starts with the familiar and builds up from there. The US renewable energy industry is fortunate to be a growing industry but still one where those "originals" know how and where to seek each other out. This coalition of the willing has for decades met at conferences, lobbied and educated through valued trade associations, and met at breakfasts and closed-door meetings to share their vision for the industry.

I credit the ability to connect with many of these original risk-takers and to pull the history and stories that powered the beginnings of the renewables industry to my time at ACORE. Under the leadership of Mike Eckhart, my first boss in renewables, I learned the importance of intellectual curiosity combined with the ever more important task of networking to build a rolodex of colleagues that would remain with me throughout my career.

While I was at ACORE, I experienced renewables' first finance, policy, and technology conferences as well as the charismatic leaders who were the firsts of their kind. Not all those at the conferences came from similar backgrounds.

Some were entrepreneurs from Silicon Valley. Others were finding their way to this new industry after failing to find a home in traditional industries such as banking. Still others were called by the inherent "goodness" of an industry they could create a career in. This industry remains one where seven degrees of separation is still unheard of. Indeed, at times it still seems like we have just one degree of separation.

As such, I went back to the beginning, pouring through old ACORE board minutes, news clippings, and memories that at times brought a hearty chuckle. Just like conducting a job search, I pulled together an initial list of the fifty leaders I believed made an invaluable impact on our industry and worked my way down to the ones I'd feature in this book.

Given my career journey in the nonprofit and for-profit sides of the business as well as a plethora of books already on the topic, I decided consciously not to feature pure technology innovators in renewables. An example being Russell S. Ohl at Bell Labs, who invented the first silicon solar cell in

1941. In addition, several leaders have already been featured in solely dedicated books about them, so I decided to avoid duplication.

To give laud to these leaders, I've referenced many of the books about them in the "Resources for the Renewables Professional" section of this book. In this section, I also provide resources that point to invaluable insights into various markets—like the distributed energy sector—as well as extensive backgrounds on renewables technologies such as the history of solar.

Those I interviewed I knew personally from previous ACORE boards while others spoke at conferences I organized or partnered on deals for teams I worked with. Reminiscent of my ACORE days when I "dialed for dollars," selling exhibit booths for PowerGen Renewable Energy and Fuels, a technology trade show that was a joint venture between ACORE and PennWell, I reconnected with all twenty-five leaders over virtual Teams sessions or through in-person interviews.

Ending each call, I asked a question that was a recommendation given to me by Mike Eckhart when approaching any new customer vertical or market leader.

"Never end a call without asking for an introduction to three other individuals they recommend," Mike once told me. Employing his recommendation, every interview ended with that question. Before I knew it, I had seventy-five more individuals to interview.

LinkedIn also proved quite useful after I placed a message back in September of 2021 asking for word-of-mouth recommendations, which shockingly resulted in more than fifty emails in one day alone. I quickly realized I would never be truly able to include everyone in my research or in this book.

Indeed, I also didn't want to write a dull book, on a topic such as the history of the modern US renewable energy industry, starting with the Carter administration and the Oil Embargo of the 1970s. We truly are blessed with so many people who were the "first of something" in renewables—enough to fill a book series on the topic.

I faced the same conundrum when parsing down the list by looking at the first US companies in the sector and the leadership that created them. Many have helped us get to the point where renewable energy investments are seen as delivering better financial returns than fossil fuels.

In fact, findings compiled by the Imperial College London and the International Energy Agency (IEA) determined that the rate of return on renewable energy investments over a five- and ten-year period yielded 200.3 percent returns versus 97.2 percent for fossil fuels.

So, after looking at a list of over one hundred individuals and corporations, and knowing I could only feature twenty-five of them, I applied a question as a lens to my research. Were these individuals true risk-takers?

This question, which I admit does not provide a definitive or scientific approach given its subjective nature, provided me with a challenge in narrowing down the list. *Merriam-Webster,* in its definition of "risk" as a verb, defines taking a risk as "incurring risk" or "exposing oneself to hazard or danger." Synonyms for risk range from "adventure" to "chance," "gamble," and "venture."

Looking at the histories of these renewable entrepreneurs and executives on my list, I began to shorten the list. Granted, I heard of many immense sacrifices made by individuals when embarking into renewables that I would not want to downplay.

These sacrifices ranged from having to weigh the decision of taking on an entrepreneurial venture with no pay—while having a young family at home—to moving cross country to take on a new job not knowing a soul.

But these were sacrifices and not risks. Having come to the realization that I would need to make a tough call and define what I assumed as risk in the context of the original risk-takers in renewables, I went ahead and drafted a definition.

A risk-taker in the early US renewable energy industry was one that: (1) made a conscious decision to enter the industry, forsaking the comfort of a previous career track, (2) experienced and survived the rollercoaster highs and lows our industry faced between the 1970 and the 2010s, (3) somehow thrived and actually grew a business or venture despite all this, and (4) contributed to the overall modern industry through a concept, technology, or platform that we all benefit from.

The fifth, and last part of my definition, was personalized to me. (5) How did they make an impact in inspiring me as a renewables professional? This last criterion helped to make each of these stories more personal and relatable, avoiding making this journey a compilation of glorified biographies that could be found on LinkedIn.

With this definition in hand, I began this journey to gather the stories behind those who took the risk and answer a variety of questions. What motivated them? Were they self-aware that they were taking a risk? What powered their vision to look past a nascent industry and see prosperity in their future? What excited, inspired and drove them to want to build out the industry?

By applying this methodology, I hope I'll do justice to both those I featured as well as the industry itself. In writing this book, I had moments of doubt whether I was the right person to embark on this journey with you all versus the talented journalists and writers we have in our industry.

With each story, and with each reflection on the traits that make up these amazing risk-takers, I found my voice. We ultimately all have the makings of a risk-taker in renewables, no matter where you are currently in your career journey.

With that, we will now begin our journey to when we as an industry had to take the risk—to be curious about how to work with tribal lands and peoples right here in the US.

PART 2

THE RISK-TAKERS

CHAPTER 5

Curiosity

Tracey LeBeau

"I took a leap based on a gut feeling. I'm curious by nature and always gravitate toward complexity and new things. But it's also all about who you know, networking, and putting it all out there," recalls Tracey.

Tracey LeBeau and I met through a mutual acquaintance, Matt Ferguson, who in 2007 became a principal and the renewable energy lead at accounting firm CohnReznick, LLP. For as long as I could remember, Matt has had an interest in tribal lands.

Having launched the new practice for the firm, Matt wanted to connect his work surrounding valuation and financing of renewables to potential collaboration with tribes in gaining them energy independence.

Through his work, he met Tracey, who commuted between Phoenix, Arizona, and Washington, DC, and had her own interests as a Native American working in renewables. They forged a professional relationship and long-standing friendship that I was fortunate to tap into. From the first day I met Tracey, her intellect and curiosity to understand how and why things worked struck me as did her dedication to civil service.

Tracey's long-standing civil service started like many of those in DC. An astute law student, Tracey excelled at her environmental law studies, graduating in the 1990s. Like many with an environmental law background, she was beckoned into the oil and gas industry.

Working mostly on the environmental reclamation side of the business, she focused on the upstream exploration production and found herself in northern Alberta just as the oil and gas boom began in Western Canada. Working with communities that were "boom towns," she assisted in setting up energy production facilities and ensuring that communities were sustainable from an economic standpoint. During this time, Tracey sensed that curiosity would empower her journey.

Curiosity has been an inherent trait many of us either capitalize on or subdue. Albert Einstein famously said, "I have no special talent. I am only passionately curious."

Israeli American Astrophysicist Dr. Mario Livio would agree with Einstein, adding that curiosity is a fundamental human trait and that everyone is by nature curious. According to Livio's book *Why? What Makes Us Curious*, the object and degree of that curiosity differ depending on the person and the situation.

"Curiosity has several kinds of flavors, and they are not driven by the same things," he adds. He defines a specific manifestation of curiosity as perceptual curiosity, and describes it as, "A bit like an itch that we need to scratch" in our consistent search for acquiring knowledge or resolution on a topic or matter.

They key word in his phrase is the addition of the word "perceptual," which signifies that the curiosity is based on the individual's perceived understanding of the environment around them, factoring into how they approach or resolve their curiosity.

Winding back to before those days in Alberta, Tracey assessed her strengths before entering the energy sector. In her mind, she was gathering all the open-ended questions around how to make boom towns more viable and energy production a more accepted part of a community's financial strategy.

Tracey foresaw a new and growing midstream market as the energy market deregulated in the early 1990s. "I wanted to understand the whole energy value chain and figured out immediately that there was one lone aspect of this value chain that I wanted to understand—and that was that energy development really brings value to a community," she notes.

She began to make inroads with executives who had left Enron and AIG on good terms, and her networking eventually led her to be recruited for a start-up in Kansas City named American Energy Solutions. The firm, set up to focus on public power creation in the US market, was the "backroom" marketing arm for a lot of cities—also known as "municipals"—for natural gas and electricity.

Public power utilities were created as divisions of local government, owned by the community, and run by boards of local officials accountable to the citizens. While each public power utility was different, "All have a common purpose: providing customers in the community with safe, reliable, not-for-profit electricity at a reasonable price while protecting the environment," according to the American Public Power Association (APPA).

Tracey was intrigued by public power and her new colleagues, who in many ways were pioneering and navigating the beginning of the retail market's takeoff. The burgeoning "new" retail energy market would allow consumers to choose among competitive suppliers and determine for themselves what energy supplier best served their home or business.

As the market began to trickle out retail deregulation, she got to work on forming natural gas aggregation groups as well as focusing on creating commercial and industrial entities beginning to wade into the market.

Tracey was in her twenties and soaking up the experience, which came with its daily ups and downs. Daily challenges included focusing on siting and permitting issues as well as diplomatically dealing with the full spectrum of those involved in energy production—ranging from pipefitters on site to energy traders.

Through it all she remembers it, "was a neat thing to experience and watch a start-up that was carving out a boutique niche in the utility and energy marketing industry."

At times, a young Tracey was the only female energy worker for miles, staying in various motels with hundreds of pipeline workers and being offered protection and safety by motel front desk staff who were concerned for her. But being the first of many women involved in the burgeoning energy sector in Alberta didn't stop her.

To Tracey, "This was part of my plan," but she admitted, "I was exposed to the roughest and humblest of the energy industry." And the rough times were indeed coming.

The California energy crisis of 2000–2001 rocked the energy sector. California found itself having a shortage of electricity due to marketing manipulations and capped retail electricity prices. However, this provided an opportunity for the energy market to be reborn, and Tracey's sense of curiosity kicked in.

Through her work in the Kansas City start-up and then later working for a Kansas Pipeline Operating Company, she began to network with a variety of other start-ups in Silicon Valley, and her passion for sustainable community development issues was recharged. Having never initially worked on community related renewables projects, she gravitated toward the renewables industry.

Joining Innovation Investments as their vice president of Wind Development in 2006, she began to work with developers and investors as well as advising a range of local governments across the US. Tracey pulled together consortiums of parties to find a way to make renewables projects work. "We chased best-in-class land, only to realize coming out to the parcel that the grid really wasn't set up there."

Curiosity came in many forms for her and her team, which in many ways was ahead of its time in trying to crack the code of electricity transmission. Daily discussions centered around out how to provide reliable energy from renewables to rural communities far from power centers.

Tracey found herself intellectually curious about the projects she was looking at, stating, "I was also trying to crack the code of how to bring people together to make renewables really happen at any kind of scale."

In cracking the code, a couple of what Tracey refers to as "aha" moments came up. The first, in 2009, was the advent of the cash grant. The Section 1603 program was created as part of the American Recovery and Reinvestment Tax Act of 2009 (ARRA) to increase investment in domestic clean energy production. Under Section 1603 the Department of the Treasury made payments—i.e., cash grants—in lieu of investment tax credits to eligible applicants for specified energy property used in a trade or business or for the production of income. The purpose of the 1603 payment was to reimburse eligible applicants for a portion of the cost of installing the specified energy property.

Specified energy technologies that were applicable for the grant included solar, wind, geothermal, biomass, fuel cells, hydropower, combined heat and power, landfill gas, municipal solid waste, and microturbines. The Section 1603 Program has disbursed over $26 billion to help fund 109,766 clean energy projects that are estimated to produce enough clean energy to power over 8.5 million homes.

The awards varied in size ranging from $180 to over $500 million and included projects located throughout the United States and the US Territories. This in many ways was a gamechanger for the US renewables industry, jump starting the development of projects and the developer industry.

The second "aha" moment for Tracey came while she was back up in Canada looking at potential natural gas projects in Nova Scotia. At that time, Hydro Quebec had put out their one thousand MW "Request for Proposals"—also known as RFPs—which sent shockwaves due to its size.

But the local content manufacturing requirement of the RFP resonated with Tracey the most. Again, her curiosity around how these projects were going to impact the local communities around the Canadian Maritimes stuck with her.

She was specifically curious to see how mandating local content manufacturing would affect local workforce development, recharge community economics, and provide a long-term vision of stability—all based on renewables deployment.

Tracey continued to grow in her career, cofounding Red Mountain Energy Partners and later becoming partner and co-chairing the Renewable Energy Practice at international law firm Dentons. During those years, she focused in on her community building passions and was "all over the place."

Tracey pulled several consortiums of stakeholders together both in Southern California, focused on large scale solar as well as up in the Dakotas, focused on a few wind developments. "And I still continued to try to crack that code on transmission," stated Tracey.

She also got involved in pulling together a consortium of the wealthiest gaming tribes on the West Coast—including tribes like the San Manuel, Agua Caliente Band of Cahuilla and a few others—with the purpose of deploying renewables.

Working with them, Tracey focused on accelerating the signing of agreements to do large scale solar development. San Manuel and others had already formed a very successful intertribal joint venture investment vehicle called Four Fires, so she and the team used that as the model. However, the tax equity piece was tough to work through to take full efficient advantage of the (then) investment tax credit (ITC).

According to Tracey, when they got right down to the end, this initiative ended up not coming together. The failure, though, fostered a "seed stuck in my head," she recollects. "As yet another code to crack."

Her attempts to continue to expand community growth around renewables as well as the overall industry's understanding of transmission did not go unnoticed.

In 2011, President Obama appointed her as the Director for the US Department of Energy's (DOE) Office of Indian Energy Policy and Programs. Her appointment was pivotal as its main focus was to establish this new office, which was authorized by statute. The office came into existence to manage, coordinate, create and facilitate programs and initiatives to encourage tribal energy and energy infrastructure development. Administratively, the office was to also coordinate, across the department—as well as the federal government— those policies, programs and initiatives involving Indian energy and energy infrastructure development.

Since she'd descended from the Cheyenne River Sioux, this was meaningful for Tracey, and she went straight to task using her community building skills. For two and a half years, her goal was to stand up the program, put it on a successful path, and find the right people to manage it.

Two specific collaborations defined her success in the short time she was at DOE. One included the Moapa Paiute in Southern Nevada's first large-scale solar development and the other focused on creating a foothold for Alaska's renewable energy sector.

The first collaboration was an intergovernmental collaboration with the Department of the Interior, which resulted in one of Indian Country's first large-scale solar developments.

Placing renewables was, "New on federal lands, much less Indian lands," affirms Tracey. "We had no way to figure out a way to evaluate or do the valuation around the proposed terms and conditions in the lease." In addition, the tribe associated with the project depended on their relationship with Interior—as the fiduciary party—to guide them since they had never done a renewables project.

The Department of the Interior needed help, and they knew who to call.

Tracey had hung up the phone and her curiosity was piqued. It was time for her team to tap into the resources of the national laboratories that DOE was known for to solve this challenge. DOE worked together with the National Renewable Energy Laboratory (NREL) and a variety of other national laboratories to model the various terms and conditions against what they saw in the market.

The collaboration that sparked a collective curiosity among the modelers and political staff involved providing the final information needed to get the project started. This curiosity from Tracey and the team drove innovation in creating the first data metrics by which to value the proposed terms and conditions around these first leases on Indian lands, providing a breakthrough for DOE.

Her ability to marshal support and resources within the department led to the second collaboration that defined Tracey's leadership at the DOE.

Alaska had been on her mind for a while as a location that was both on the leading edge of experiencing climate change and, with the ascent of Senator Lisa Murkowski, a vocal advocate for the energy industry. She had a hunch and saw an opportunity to do the right thing.

Again, curiosity played a role, coupled with the insight Tracey had garnered during her previous collaboration and resulted in action. Leveraging relationships and resources, she was able to partner with the Denali Commission and the Alaskan energy state agency to re-energize a stagnant DOE office that had lain empty for a better part of a decade.

Using her experience in community building, she saw that she would be able to revitalize several traditional energy boom-town communities through renewable energy deployment.

The one thing she knew from observing best practices was that throwing money at communities to get projects constructed often left her with projects that were not well-maintained and often un-operational. In the past, many of these investments went to rural areas where "They couldn't pick up the phone and have someone come out to fix a wind turbine," notes Tracey.

She saw this as a missed opportunity that needed to be remedied. The newly standing DOE office in Alaska would focus not only on the deployment of assets but also on setting up the community ecosystem that would ignite workforce development. This in turn would help operate and maintain energy assets, building an economically sustained community.

Given that Tracey had control over grant funding and harking back to Hydro Quebec's revolutionary stance on local content manufacturing, she remained steadfast on the community-building requirement of any project the office would engage. She was not going to budge on the requirement and the potential to revitalize local workforce and community economies.

"I said, okay, we're going to focus in on capacity building and make sure to also bring to the community not only financial assistance but pair it with technical assistance and community strategic planning," recollects Tracey. "I really felt I was going to make a difference for the longer-term commitment of a community toward development of energy projects."

She wanted the community to see themselves in the promise of these infrastructure projects and not feel like pass-through communities that have no say nor any access to the opportunities such developments possess. Tracey knew, "If they feel like a pass-through community, that's where you get the resistance."

Having a community see itself as a beneficiary of infrastructure development was no easy task and required focusing on how new energy technologies could provide invaluable workforce and education training, steady paying jobs, and financial stability upon which communities can thrive and grow.

Tracey went on to lead the Office of Indian Energy Policy and Programs and to inaugurate the START program. The Strategic Technical Assistance Response Team Program assisted tribes in the forty-eight contiguous states and Alaska with renewable energy and energy efficiency projects. Through START, technical experts from the US Department of Energy Office of Indian Energy and DOE Laboratories coordinated with tribal leadership and staff to work toward achieving project goals.

"I felt that if we were going to really push for infrastructure deployment and do this at scale, we needed to get commitment from all the states and to solve the community involvement issue," stated Tracey.

START ended up supporting over thirty Tribes and Alaska Native villages through more than $100 million dollars in direct renewable energy project investments, helping drive energy and energy efficiency closer to implementation on tribal lands from Arizona to Wisconsin and California to Alaska

Tracey's continued quest for answers gave her curiosity a function in her career journey. Psychologist Dr. George Loewenstein suggests that a small amount of information serves as a priming dose that greatly increases curiosity. Fellow psychologist Dr. M.J. Kang, along with a number of colleagues, found through their research that curiosity enhances learning, which is consistent with the theory that the primary function of curiosity is to facilitate learning.

At the end of the day, curiosity is all about acquiring knowledge and, in an evolutionary sense, helps to function as a driver of evolution for both individuals and the renewable energy industry alike. As much as curiosity was a catalyst, the journeys Tracey took as part of her quest provided a strategic benefit and ability to overcome perceived industry barriers while seeking out paths and strategies not yet explored.

In simple terms, curiosity provided a launching pad to bigger and better ideas, providing Tracey an opportunity to focus on current challenges while at the same time thinking beyond to the next role she would play.

That next role came in 2021 when Tracey became the Administrator and Chief Executive Officer at the Western Area Power Administration "WAPA". WAPA is one of four power marketing administrations within the DOE whose role is to market and transmit wholesale electricity from multi-use water projects.

With a service area encompassing a fifteen-state region of the central and western US, WAPA oversees a more than seventeen-thousand-circuit-mile transmission system that carries electricity from fifty-seven hydropower plants with an installed capacity of 10,504 megawatts.

True to Tracey's passion, WAPA sells power to preference customers such as federal and state agencies, cities and towns, rural electric cooperatives, public utility districts, irrigation districts and Native American tribes. They, in turn, provide retail electric service to millions of consumers in the West.

Looking back, "There were many brilliant folks along the way in this journey," acknowledges Tracey—from Arun Majumdar, the Founding Director of the Advanced Research Projects Agency-Energy (ARPA-E) in 2009 at the Department of Energy to Julia Prochnik, the former Director of

Western Renewable Grid Planning at the Natural Resources Defense Council (NRDC). Many of these leaders helped to ignite the curiosity feeding the insight that propelled Tracey throughout her career. They also provided a much-needed oak tree to reflect on leadership in our industry.

These types of conversations make Tracey reflective. "We're moving a million miles a minute and don't present the bigger picture that we're still working toward. It's the age-old problem with leaders that we get so busy doing what we're doing that we don't take the time to stop to talk about what we're doing."

Curiosity, combined with strong emotional intelligence and dedication to the communities she served, gave Tracey the initial combustion she needed to propel through her journey. The same curiosity continues to offer us the fuel to get the mission done by delivering the message that communities can be active participants in their own successes through the adoption of renewables.

CHAPTER 6

Persistence

Scott Sklar

"The stars had aligned, and it was a big bang in its early stage. Those of us who went through it are all close friends. It was like being in the military in a war. We were in a war of ideas, a war of energy access, and a war for energy democracy," begins Scott.

Scott Sklar is no stranger to the Washington, DC, scene. Since the 1970s, he has been roaming the halls of Congress and K Street as a missionary for renewables. He was known for connecting big industry ideas to the politicking that was necessary for passage of everything from the Clean Air Act and Clean Water Act.

Scott's office for decades served as the birthplace of many trade association leaders and other politicos who continue to dominate renewables conversations nationally. Scott and I knew each other from working together on a number of educational policy events on the Hill. In my early days in

DC, he had always opened the door to me for any policy questions I had.

When you ask anyone in the industry what trait is best associated with the name Scott Sklar, many jump right away to say, "He is persistent." Perhaps because I can relate to being stubborn as a productive narcissist, I've always felt that stubbornness—when tied to a cause or desired outcome—can be a benefit.

Manfred F.R. Kets de Vries, a clinical professor of Leadership Development and Organizational Change at INSEAD states, "Stubbornness makes us persevere. It helps us stand our ground when everyone else is trying to tell us that we are wrong. Used with discernment, stubbornness can be a strong leadership quality and a key determinant of success. Because stubborn people know what they want, they tend to be more decisive."

Anyone who is familiar with DC knows that many of the best and most impact-filled strategies occur over breakfasts, usually at the Army Navy Club or Old Ebbitt Grill.

Those early conversations that were held between initial renewables pioneers in the late 1990s and early 2000s charted the industry's course and ultimate destiny. One constant at almost all the breakfasts was Scott's participation.

Like many before, and indeed, many after, he came to Washington during the Carter Administration, wanting to make a change. He began his fifty-plus-year career working in the 1970s as a Senate Aide for then Senator Jacob Javits (R-New York), who was a senior senator. Put in charge at first to cover military issues, things changed suddenly for Scott at the onset of the oil embargo with OPEC cutting 3 percent of gas exports to the US.

At that time, there were little to no energy staff on Capitol Hill to cover the issues. Of the staff that did exist, most were tied to committees and working on nuclear issues, public lands, or offshore drilling. Scott was in many ways a singular voice and an immediate expert on energy who persistently wanted to apply his knowledge on renewables to the many policy conversations he was engaging in.

Having come out fresh from graduate school with experience dealing in small solar, micro-wind, and hydropower water turbines, Scott's first assignment covering military issues for Senator Javits exposed him to topics ranging from reliability and resiliency to efficient deployment of energy.

The situation presented an opportunity in the mind of Senator Javits. He had always appreciated Scott's persistence and, though he had little interest in energy topics, he knew Scott had a passion in them. Although he was a Republican Senator coming from New York City, Senator Javits knew that upstate New Yorkers—especially farmers—had an interest in developing alcohol fuels and solar energy.

The senator was perpetually stunned by the amount of correspondence he got from his constituency on the topic. Competition to jump on this new topic was fierce among the staffers from the New York State delegation, but within a few weeks of the Oil Embargo in the late seventies, Scott became the newly anointed energy aide for the senator.

As staff on the Hill was figuring out how to cover the situation on our national energy strategy, Americans were gathering in long gas lines across the US. According to Scott, this was the "First time Americans had experienced an energy shortage of this kind since the discovery of oil and gas in Titusville, Pennsylvania, in 1859."

Combined with the commercialization of coal and the American discovery of nuclear energy a few decades later, Americans were used to plentiful energy. For the first time, many Americans realized most of what they relied on was indeed not home grown. The times called for people to come together, and Scott saw that.

Seeking allies across DC—and indeed across the US—a "coalition of the willing," which was the first of many for the burgeoning US renewable energy industry, began to form. Many joined Scott in strategizing what recommendations to make to senior officials in Washington.

First came Denis Hayes, who led Worldwatch Institute, a globally focused environmental research organization based in DC. It was founded by Lester Brown, one of the world's most known environmentalists. Denis later helped to create the first ever Earth Day.

Also joining them was a former military leader, Bill Holmberg, who was at the EPA overseeing alcohol biofuels work and was about to move over to the Energy and Research Development Administration, the pre-cursor to today's US Department of Energy. David Morris, a famous author and vice president of the Institute for Local Self-Reliance also folded into the group along with Harry Barlow.

Reflecting on the oak tree metaphor, Scott recollects that, "We all gravitated to each other because, frankly, we were lonely. We were all intellectually, emotionally and socially going on this parallel journey together."

Gathering recommendations from the group, Scott knew he had to forge alliances to get the light of day on what he considered a comprehensive energy plan, which he hoped would give renewables the necessary foundation to build from.

Scott made a decision to arrange a meeting with Proctor Jones, the staff director who oversaw Energy and Water Development for the Senate Appropriations Committee. Proctor had been on the Hill for thirty-five years and now worked for Senator Bennett Johnston, who at the time was a Democrat representing the state of Louisiana—a large oil proponent.

Both Proctor and Senator Johnston listened to Scott as he read off his list of recommendations, which included significant changes to how the Energy and Research Development Administration functioned. These recommendations ranged from how government could support the growth of renewables through incentives to ways to support American entrepreneurs who already had renewables technologies ready to go.

Smiling at Scott, the senator got up, walked over to him, and put his arms on Scott's shoulders.

"Scott, I just want to tell you, and I'm very serious in saying this. I do like your enthusiasm, and I'm going to do many of the things you requested as a friend, but I have to tell you, there will never be wind farms. There will never be acres and acres of solar. You're fighting against a six-trillion-dollar industry and your entire renewable industry is two hundred and fifty million dollars, of which 95 percent is hydro. Everything you're talking to me about is twenty times more expensive. It's just never going to happen."

And with that, and to the shock of everyone around, the senator gave Scott most of what was on his list. His persistence was paying off.

This win empowered Scott to create the first Congressional Solar Caucus, which was completely bipartisan, including help from Terry Johnson, who was the Staff Director for a then Senator Gary Hart (D-Colorado) as well as Dave Springer who worked for then Senator Howard Metzenbaum (D-Ohio). "So, we were all teamed up as an empowered group of staffers with a bipartisan cause to advance solar," Scott says.

The caucus started up education programs and teamed up on legislation over the next three years. Given the size of the renewables industry, both the oil and nuclear groups on the Hill ignored the caucus, though they did, "Beat the crap out of us once in a while."

By not participating in efforts and advocating for others not to participate, these anti-renewables coalitions proclaimed that solar had a small part to play in any meaningful energy legislation. Scott persisted in his leadership and maintained his belief in the caucus.

The group nonetheless continued undeterred, building up educational campaigns and promoting the concept of a viable renewables industry on Capitol Hill. Partnering with the "coalition of the willing" that Scott had gathered of environment and NGO leaders, they tapped into nationwide networks of renewable supporters staging numerous national events.

These events ranged from organizing alcohol fuel caravans from Montana to Washington, DC, to showcasing actor Charlton Heston's alcohol powered Corvette as part of a commemoration of the then fifth anniversary of the Oil Embargo in 1979.

"And at some point during all this, we got a lot of legislation through," notes Scott. This legislation included the creation of today's US Department of Energy (DOE) as well as beginning conversations on what would become the Clean Air Act and Clean Power Act.

Following on the heels of this legislation was another game changer for renewables—the passage of the Public Utility Regulatory Policies Act known to many as "PURPA"—which was passed as part of the National Energy Act and enacted on November 9, 1978.

The act promoted energy conservation through a reduction on demand and promoted greater use of domestic energy and supply of renewable energy. For this first time, private power producers would be allowed to hook to the grid, paving the way for many of the renewables developers of today.

Scott's popularity had risen on the Hill as one of the "go to" experts on renewables, but he felt the time had come to take on a new challenge and started to have conversations with close colleagues. Then one afternoon in the early eighties, he received a call that once again would change his trajectory.

The Chief of Staff from the Office of Democratic Majority Leader Mike Mansfield (D-Montana) was on the other end of the line and asked Scott to come in to meet with him. For those of you who are political history fans, this name should ring a few bells. Following the Watergate burglary in 1972, Senator Mansfield set up the Watergate Committee, chaired by Senator Sam Ervin (D-North Carolina).

To help reassert congressional authority, Senator Mansfield cosponsored the War Powers Act of 1973. Under his leadership in the 1970s, the Senate also adopted a series of institutional reforms, from the "sunshine" requirements for opening committee meetings to public scrutiny to reducing the number of votes needed to invoke cloture from two-thirds to three-fifths.

Scott walked into Senator Mansfield's office and sat as he completed a call, thinking about why he was there and what the "political ask" would be from the senator.

Senator Mansfield got off the phone and said, "I hear you're leaving Senator Javits."

Scott responded, "How do you know that?"

"I'm the majority leader," the Senator answered.

The senator continued stating that he had worked to put together language in an appropriations bill in the late 1970s with EF Schumacher that helped launch a National Center for Appropriate Technology and Applied Laboratory in Montana in 1976. Scott knew the name well.

E.F. Schumacher was a German-British statistician and economist best known for his proposals for human-scale, decentralized and "appropriate technologies." He served as Chief Economic Advisor to the British National Coal Board from 1950 to 1970 and founded the Intermediate Technology Development Group (now known as Practical Action) in 1966. In 1973, he wrote a book titled *Small Is Beautiful: Economics as if People Mattered,* which was ranked by The Times Literary Supplement as one of the one hundred most influential books published since World War II.

"Scott, we'd like you to lead the Washington, DC, office." The senator was then summoned to a meeting and rushed out of his office with a smile.

Scott accepted the position, launching the laboratory's DC office and then spent a year in Montana in 1981 overseeing a combination of sixty-eight NASA scientists and a few individuals from the Peace Corps who were looking to implement renewables in rural communities.

Scott was impressed with the team. "These were all the guys doing advanced stuff with satellite solar in space....and the Peace Corps guys were doing some innovative on-the-ground experimentation."

Having been at the laboratory for a short time, he was surprised to receive a call from Senator Gary Hart (D-Colorado), who was in the process of setting up the Solar Energy

Research Institute, otherwise known as "SERI," which to many today is known as the National Renewable Energy Laboratory (NREL) in Golden, Colorado.

On the line also was his staffer, Terry Johnson, an old friend from the Congressional Solar Caucus. At that time, Denis Hayes from Scott's "coalition of the willing" was also at SERI. Gary, Terry and Denis had a mission for Scott.

Nine environmental groups—ranging from the Sierra Club and the Union of Concerned Scientists to the Environmental Defense Fund—were going to announce the creation of a solar advocacy group in 1978, called the "Solar Lobby" to complement the work of an informal trade association that was formed in 1974. We know that association today as the Solar Energy Industries Association (SEIA).

The Solar Lobby worked hard to counteract the tumultuous times of the early 1980s, which saw the Reagan Administration in office, and ushered in a mounting fear that the milestones made to date would be swept away. Unfortunately, these fears were proven right.

President Reagan is notorious in the renewables industry for one emblematic incident in 1986, in which he removed solar panels the Carter Administration had previously installed on the White House roof.

Reflecting on this period, author Natalie Goldstein wrote in her book, *Global Warming*:

"Reagan's political philosophy viewed the free market as the best arbiter of what was good for the country. Corporate self-interest, he felt, would steer the country in the right direction."

The Reagan Administration moved forward with drastically cutting the newly inaugurated DOE's funding for research and development in renewables, keeping in line with President Reagan's comments on the issue during presidential election debates with President Carter. In addition, President Reagan signed a tax bill known as the Economic Recovery Tax Act of 1981 (ERTA), or Kemp-Roth Tax Cut, that severely cut federal renewable energy research funding and residential tax credits.

Given what was occurring, the SERI trio encouraged Scott to come in as a Political Director for the Solar Lobby and bring together what was a disparate group of solar constituencies into this newly formed solar advocacy group.

The industry was primarily made up of disparate suppliers of glass, steel, and various other components used in solar module manufacturing. Industry players were hired by various entities to produce "bespoke" projects—for satellites, for example—with mass-produced solar being completely unheard of.

Most of the industry was focused on solar heating versus the overwhelming solar PV industry we have today. Scott was persistent and persuaded the various factions to come together and be a much louder, unified voice that would present a modern solar energy industry spanning solar PV, solar-thermal and solar-water heating sectors.

Throughout the eighties, Scott testified with this unified group of members at numerous hearings on national security, diversification of energy resources, and the need for energy efficiency and renewable energy. He built a direct conduit with then Energy Secretary nominee John Herrington. That resulted in an open exchange of renewables data and information with the DOE, much of which was opposed to by the administration but accepted and utilized by the Secretary.

Despite the political climate, the Solar Lobby achieved its objectives and worked together with SEIA to promote ongoing research and development funding, which kept solar energy a priority for the DOE.

Scott's work on the behalf of the Solar Lobby got him immense laud, and shortly after in the mid-1980s, the National Biomass Industry Association asked Scott to also be their Political Director, with the goal of increasing the use of biomass power and creating new jobs and opportunities in the biomass industry across the US.

Putting the growth of SEIA and the National Biomass Industry Association in perspective, the spur around the mobilization of the renewables industry resulted in the concurrent birth of the additional renewables trade associations in DC spanning the early 1970s into the 1980s.

During this time, in addition to SEIA and the National Biomass Industry Association, the National Hydropower Association (NHA) was formed alongside the American Wind Energy Association (AWEA) and the Renewable Fuels Association (RFA). Trade Association "legends" took command, ranging from Rhone Resch and Dave Hallberg to Randy Swisher and Linda Church-Ciocci—many of whose industry groups started out of Scott's office.

As all this was going on in Washington, Scott and his "coalition of the willing" directly helped in setting up an individual membership Colorado-based organization to support solar called the American Solar Energy Society (ASES), which over the years ballooned to over thirty thousand members.

The late 1980s into the 1990s brought many new solar market entrants, especially from the military and defense sectors, including the likes of Rockwell International and McDonnell Douglas.

The mainstream energy sector was also starting to look at the industry, including BP, who had decided to get in the solar panel manufacturing business by acquiring Solarex in Maryland. Slowly but surely, the industry persisted, pulled up its bootstraps, and began to manufacture panels in the US.

This all did not happen overnight. It came out of a call for standardization from large corporations and developer offtakers looking to incorporate a variety of solar technologies into their offerings.

When Scott spoke with his large corporate members, they characterized their needs by stating that, "We don't buy refrigerator compressors. We buy refrigerators. We don't buy car engines. We buy cars. We want standardized modular systems that ultimately we can finance through service contracts. Everything your industries are making is custom, and it cannot grow that way."

This call was ultimately heard by not only the US solar industry, but by the global industry, with many industry entities in the 1990s and 2000s beginning to develop more standardized approaches to manufacturing and product development. In this regard some claim the US lost its advantage to the Chinese who entered the market in the early 2000s, and in the span of a decade, they more quickly and efficiently produced standardized solar panels.

Over the course of the 2000s, and his next fifteen years of working with a variety of the renewables trades, Scott encountered every political dialogue and topic you could imagine in Washington due to a maturing solar industry. From tax credits, zoning and interconnection codes to codes and standards debates—with electrical contractors and buildings code enforcers—Scott was sought after for his advice.

And just when you'd think he'd had enough of helping to set up advocacy and educational nonprofits, Scott was involved in a collaboration with the natural gas industry to create the Business Council for Sustainable Energy (BCSE) during the administration of George H.W. Bush.

He simultaneously worked closely with Carol Werner to set up the Environmental and Energy Study Institute (EESI). Scott also served as an Advisor to Mike Eckhart in his initial days setting up the American Council on Renewable Energy (ACORE) alongside vested stakeholders that included Judy Siegel and Frank Tugwell from Winrock International.

Scott's journey reinforces in many ways writer Harvey Deutschendorf's theory in a piece he did for *Fast Company* magazine. In his article, he claims that highly persistent people have seven habits that empower them to achieve success. Namely: (1) an all-consuming vision, (2) a burning desire, (3) inner confidence, (4) highly developed habits, (5) the ability to adjust and adapt, (6) commitment to life-long learning, and (7) role models who act as guides and mentors.

Scott and many early renewables risk-takers personified these habits. They persisted even when faced with fluctuating energy policies and early-day industry supply-chain issues. Despite these and other major setbacks, they continued forward, figuring out ways to maneuver around challenges to achieve strategic successes and inching closer to their ultimate goals.

From his early days setting up the Congressional Solar Caucus, Scott had an ability to confidently translate and pitch his vision of a renewables-filled future to politicians.

His never-ending quest to adapt to the political environment and keep on top of evolving solar technologies coupled with persistence resulted in his ultimate goal—giving birth to a strong US solar industry.

When asked how he prioritized all that was in front of the solar industry in the early days, he admitted that, "It was just chaos." Despite the chaos, Scott simply carried on until solutions could be found, even if the path wasn't clear. He attributes surviving those days to the close connection to his coalition of the willing:

"We all worked twenty hours a day and we were all calling each other. It was a dynamic time really. We all just helped each other out."

Scott's key to success underscores the winning tenets of persistence:

"Play to what you'd like to do and what fits your personality, not what you think you want. Take risks when you're younger and jump fully in—not because you won't be able to do it when you're older—but because you might score big or you might crumble, but if you crumble only you crumble, not your ambitions. Either way, youth allows you to build on your successes or failures."

On being asked about traits that are associated with risk-takers, he recollected his early days as a Hill staffer and tied it all back to being persistent. "We were getting the shit beat out of us most of the time, so you had to be a very stubborn person, and you had to be very committed. And you really had to not be the kind of person who worried what other people thought. A lot of the people I mentioned, are all that way, and that's why we succeeded, together."

CHAPTER 7

Connector

Tony Clifford

"I'll just be frank with you. There were enough tree huggers around Washington, DC, to make solar a viable business without incentives because we didn't really have incentives back in 2006. The 30 percent ITC had to be approved and reapproved on an annual basis. And the state level stuff was really just starting and there wasn't a whole lot going on. So, we all collectively took the risk," recollects Tony.

Tony Clifford had been part of the original solar movement, having joined Solarex straight after finishing business school at UVA's Darden School in 1975. Having pitched the first ever photovoltaic market studies to the US Department of Energy (DOE) in 1977, he completed the study in 1978 under a subcontract from BDM Corporation.

Wrapping up his time at Solarex around 1981, he stepped away from solar to become a full-time energy business consultant. Canvassing his connections, his practice was jump-started when he received a contract from Bill Holmberg, who was running DOE's Office of Alcohol Fuels. The contract resulted in the creation of a mini-book that was a little over one hundred pages, entitled *Tax-Advantage Investing in Renewable Energy*.

Initially consulting for developers of alcohol fuels projects in California, Nebraska, Iowa, and Maryland, Tony grew his practice and connected with early-stage equity financing players. He also formed connections with those in the technology research and development (R&D) space focused on technologies ranging from biomedical diagnostics and ground-probing radar, acoustic/RF devices to additional classified technologies. Tony was a full-time consultant until Congress changed tax laws in 1986.

From 1986 through 1991, he worked with several technology-based companies. Starting out at ENSCO/SecTech, which grew out of a R&D partnership he had financed and eventually sold to Wackenhut Corporation, he continued to Quantex Corporation, an optoelectronics company founded by Dr. Joseph Lindmayer of Solarex, and then to Optex Corp, a spinoff of Quantex, which was developing a novel optical storage media.

Since my earliest days in DC, Tony and his wife Ellen have been one of the most gracious hosts in the industry, opening their home to me and countless others for dinners and mentorship conversations about the ever-evolving solar sector. I claimed Tony knew everyone who was anyone and who, through his generosity of time, would assist those newbies in the sector with introductions to other leaders who could serve as mentors.

He has a genuine and direct manner of speaking, offering unbridled honesty, but at the same time, is a listener who keeps you in the center of his attention.

Tony always had an uncanny ability to connect with people and share his passion, which focused on maintaining a concerted effort inside the Beltway to even the playing field for solar from a policy perspective. "It's a true passion of mine to catalyze growth for promising solar companies," he's cited as saying when he joined the Board of DC-based New Columbia Solar.

Tony's been a connector since his first day in the solar sector. Malcolm Gladwell, author of *The Tipping Point*, is claimed by a few journalists ranging from the *New York Times*' Rachel Donadio to *MacLean's* Andrew Potter to be the first person who popularized the term "connectors."

In his description of them, Gladwell stated that they are people with a truly extraordinary knack of making friends and acquaintances. Their ability to connect with others allows them to span, "many different worlds, subcultures and niches."

He goes on to associate traits common to connectors, naming key ones as being: (1) energy, (2) insatiable curiosity, and (3) a willingness to take chances. In addition, connectors will always claim that connecting is not the same as networking.

The seasoned connector should be seen as different from a networker. As networking is often viewed as self-centered and a means to an end, connecting is driven by a genuine interest in people and purposeful engagement in bringing others together to help each other regardless of personal benefit. Connectors have the magic to connect on a deep level, open up unassuming dialogue, and connote a sense of trust that establishes a relationship.

Tony's days as a consultant, canvassing up-and-coming technologies in the optical storage sector had paid off, and he spent the next two decades serving as CEO and CFO to several other high-growth technology companies.

In 1991, as chairman and CFO, he co-led a leveraged buy-out of an optical storage device company from Nakamichi Corporation called MOST, Inc, which was a major supplier of large-format optical drives to Hewlett-Packard. The buyout aimed to raise additional capital to develop a smaller drive for the burgeoning PC marketplace.

Tony's network of connections in the optical drive sector continued to expand and he had entered into a strategic partnership with Nikon to develop a rewritable CD drive. After a successful development effort, Nikon acquired MOST in 1995.

At the behest of VC investors in his network, he went back to Optex as a consultant and then became president in an effort to complete the development of their novel optical drive. He worked hard to secure a strategic partnership with HP.

However, within a year, HP discovered a competing stealth technology in Japan that would beat Tony's company to the market by twelve to eighteen months. Seeing the writing on the wall, he then conducted a sale of the assets on behalf of his investors.

From 1997 until 2005, Tony leveraged his now impressive rolodex of contacts and decided to go back into consulting, first as a consultant to RealNet Learn Services, a video streaming platform, and then to Witten Technologies, Inc., a company using ground-probing radar to provide geophysical mapping services to utilities and government agencies.

In 2005, he was approached by an industry colleague named Neville Williams, who had gathered seed capital to form a solar company. Tony knew Neville well from the energy sector in DC and Neville himself was well-connected and had made a name for himself around the promotion of solar globally.

Neville, who had worked on focusing on deployment of solar photovoltaic panels throughout villages in India, had decided to look back toward the states and apply his "know how" in the US. His track record and passion for solar spoke for itself, as well as his connections to the "who's who" in DC working in global solar development.

He had previously founded the Solar Electric Light Fund (SELF) in 1996, a nonprofit organization based in DC, that assisted governments and community organizations to develop solar rural electrification projects in India, China, Sri Lanka, Nepal, Vietnam, Brazil, Tanzania, Uganda, South Africa, Indonesia, and the Solomon Islands. It still exists today, and focuses on efforts in Haiti, West Africa, and Colombia.

In 1996, Neville went on to start the Solar Electric Light Company (SELCO), and with cofounder Dr. Harish Handy, launched SELCO-India in Bangalore. SELCO-India, the first commercial venture of its kind in India, still exists today and has installed over half a million solar home lighting systems.

Oil neared sixty-five dollars a barrel in 2005 and was only going to increase, so Tony thought to himself that, *Maybe it's time to look at solar again.* After agreeing to join the venture, Neville brought on Tony to help define the vision and the business model for this new venture.

They successfully launched their solar development company that year and named it Standard Solar, locating the company in Maryland. With three other colleagues—Lee Bristol, Matt Griffith, and Andrew Truitt—they began to develop residential onsite solar solutions. Tony noted specific hardships faced during those early days starting up Standard Solar.

"When we were getting started, I think Lee took the biggest risk because he was an IT guy. He had this interest in solar, but he also had a family. He was doing Standard Solar on a weekend basis," Tony recollects.

Solar was a new industry with nonexistent workforce development opportunities. "People didn't know anything about solar for the most part." However, the industry had momentum and, utilizing his connector skill set, Tony sourced, hired, and trained a number of installers.

"A lot of enthusiastic people wanted to get into solar, but finding applicants who were comfortable working on roofs and had some sort of basic construction skills was tough," Tony notes. "We had a lot of education to do, which was one of the primary challenges of new industry in those days."

Tony, through his hiring and training of these solar technicians, in many ways created a connector network. In creating an oak tree of his own, he connected individuals who—through their newfound solar installation skill sets—formed a first-of-a-kind community of solar professionals in the Mid-Atlantic. All of them were dedicated to growing out the solar industry in the region.

Eventually stepping into the role of CEO of Standard Solar in 2007, Tony and his team embarked on the strategy of connecting with "tree huggers" throughout the DMV region.

According to Tony, "Our initial theory proved to be correct. A lot of people were interested in solar due to environmental concerns, but a lot of people—some of them were ex-military and people in the government— had just had it with the Middle East. They also wanted America to be energy independent."

Ranging from military generals to employees from the Environmental Protection Agency (EPA), Standard Solar began to install solar throughout the region. His ever-increasing customer rolodex and strong connections ultimately led to meeting another notable name in our industry—R. James "Jim" Woolsey, who was increasingly encouraging his neighbors to adopt solar on their homes.

Jim was primarily known for his service to the US as the former Director of the Central Intelligence Agency (CIA) as well as for being a distinguished admiral in the Navy. I had the privilege of seeing Jim often in the office at ACORE due to his friendship with my colleague, Bill Holmberg. Jim was one of the earliest adopters of residential solar and EVs. He started driving one of the first Toyota Priuses I had ever seen in DC around 2005.

As we were discussing what we needed to advance the residential solar industry in the Mid-Atlantic, Tony leaned over to Jim and asked, "What motivated you to support solar and put it on your home?"

Without a pause, Jim looked Tony in the eye and said, "Thirty years of dealing with the goddamn Middle East."

Within the first five years of existence, Standard Solar became an Inc. 500 company, with Tony attributing the company's growth to the role of government, which had a role to play in the solar industry's renaissance.

Starting in the 1970s, through the funding of research behind initial solar technologies created in the US and incentives like the ITC, he saw the industry need successful public-private partnerships to scale.

"I understood the importance of government in initially funding virtually everything related to solar in the 1970s, to show that solar could compete against a diesel generator effectively," comments Tony.

His uncanny ability to connect and navigate through conversations being had in and outside the halls of Congress complemented his hunger and ability to scope out an opportunity. "We did a whole bunch of demonstration projects back in the pork barrel days."

In this instance, Tony referred to "pork barreling" as the often-implemented legislators' practice in politics of slipping funding for a local project into a larger appropriations bill.

"I remember doing a large project at Mississippi County Community College in Arkansas, and that was because a very influential senator from Arkansas was there at the time. And we did one in Mississippi because of Senator Jamie Witten," said Tony in recollecting a project they had success pitching.

Adding to this cadre of projects—and one that Tony is especially proud of—was the first installation of solar in the District of Columbia in 1982. Having worked with Georgetown University's facilities team, he had the opportunity to present a plan to clad the newly planned Bunn Intercultural Center (ICC) with a 250 kW worth of rooftop solar, which was unheard of at that time. Tony had the opportunity to present the proposal directly to the president of Georgetown, Fr. Timothy Healy, S.J.

"We took it in to show him the proposal, and he looked at it for ten seconds and said, 'This is a very fine-looking proposal, son,'" remembers Tony.

Fr. Healy then told him, "I'll share it with my tennis partner after six o'clock mass on Sunday."

Tony was left confused and wondering, *What the hell is this?*

It turned out that Fr. Healy's tennis partner was Tip O'Neill—the Speaker of the US House of Representatives. Six months later, there was a line item in the Department of Energy (DOE) budget for a solar demonstration project at Georgetown.

According to Tony, leveraging one's political connections through unofficial lines was the way things got done back then. Connections had once more proven to be handy for Tony, who in this instance alone had exhibited three types of connector traits.

According to psychotherapist and author of *Zen Master* Mary Jaksch, connectors can't be lumped into one category. Three types of individuals or personalities are gifted with the ability to connect: (1) the traditional connector, (2) the maven, and (3) the salesman. Three subtle yet distinctly empowering traits differentiate the three.

The connector genuinely enjoys others and can relate to the conversation or situation they find themselves in. They are good at "knowing" others and are impeccable at keeping in touch with others. The maven always seems to be on top of the latest market trends and industry opportunities. People are attracted to remain in touch with them because they freely share their knowledge, just as an educator would. The salesman has the ability to build rapport while having a keen sense of being able to diagnose challenges and charismatically "selling" solutions to those challenges.

In challenging Jaksch, and specifically in the case of Tony and the early connectors in the renewable energy industry, I'd offer up our industry's early connectors had elements of all three of these personalities due to a few factors.

Namely, given resistance to initial commercialization of solar in the 1980s, the connector personality was vital in bringing together the "coalition of the willing," building oak trees necessary to gather industry knowledge, financing, and policy vision.

Tony exhibited that trait, having joined forces with Scott Sklar to plot policies that were needed to help grease the wheels of this new industry.

Second, the maven personality was found in individuals throughout the industry, gathering analytical data and technology know-how vitally needed by initial financers to get the industry going. In this instance, Tony again demonstrated this skill, having worked on the first PV study in the industry and providing vital benchmarking and data points showing the potential for solar to grow. This growth was intrinsically tied to the structuring of public-private partnerships (PPPs) between the US government and the US solar industry.

Last, the industry needed the salesman, who could pitch and give visionary and charismatic momentum to get the industry launched. Evident through the story regarding Fr. Healy, as well as countless others, Tony had the charisma needed to not only sell the solar product at hand, but also to sell a vision for the greater role sustainability would play in any business operation.

As an industry, the renewables industry was unique in that it had—and continues to have—a number of these individuals who embodied all three of these personality traits and were willing to take risks. They gravitated and connected with those who shared a similar vision for expanding the industry, forming strong oak trees.

Tony ended his conversation with me by reflecting from a risk standpoint, "I think the biggest risk was just getting the business started because solar in 2006 was still a very unproven thing worldwide, but especially in the United States."

Thanks to Tony, his risk in starting Standard Solar proved the viable solar market in the residential and commercial sectors and provided a prototype business model that would lead to the establishment of several companies that defined the US solar industry.

CHAPTER 8

Audacity

Michael Eckhart

This is your task.

This is my task.

This is our task.

On March 4, 2008, Mike had opened the Washington International Renewable Energy Conference (WIREC). His speech was the culmination of seven years building up the American Council on Renewable Energy (ACORE) and its mission of uniting the fragmented renewable energy industry. His speech was seen as a crowning achievement for ACORE, which to that point had been a small, nimble, well-oiled 501(c)(3) dedicated to the success of renewables in the US.

Mike had known for years that none of us alone could give monumental thrust to the rocket ship known as the renewable energy industry. We had a need to create an oak tree by which to gather like-minded people who believed in the industry, were persistent to succeed, and were willing to take a risk to get the industry into orbit and into the mainstream of society. As a marketer, I had signed on for this mission when I found myself standing there listening to Mike on the side of the stage at WIREC.

In closing out his speech, Mike ended by saying, "The reason we came to WIREC—the reason you came to WIREC—is that you and I felt, in our gut, that we cannot do this alone. Our only chance is to do this together."

To him, it was also important for those renewable entrepreneurs, statesmen, and investors present at WIREC to, "Have your own work recognized for its excellence and go home more confident to do what you can do to cause your country, or your company, to develop and adopt renewable energy and make it the reality of our future, indeed, the reality of our time."

I remember that moment, standing there at the main ballroom at the Walter E. Washington Convention Center and watching all the faces of those two thousand invite-only guests who were there to represent their countries' commitment to renewables react to that statement with thunderous applause.

Downstairs, another eight thousand attendees were walking through a series of exhibit halls and a maze of conference rooms, where they brainstormed and exchanged ideas on the technological, finance, and policy mechanisms that were working to scale renewables.

It was a defining week for me, and the small but mighty team at ACORE, having raised $6 million in sponsorship funding over the span of the previous year and putting on a world-class event the US has not seen since then.

WIREC 2008 was the culmination of years of momentum started by the late Hermann Scheer of Germany. Hermann was known for his impact on the development of the feed-in tariff "FIT" in the late 1990s as well as his formation of the World Council on Renewable Energy (WCRE) in 2001.

A reoccurring theme with many of the early entrepreneurs in renewables present at WIREC was a general call to action. To many, this call to action was tied to social good versus advancing a product. Much like the momentum needed to launch a rocket into orbit, you also needed a healthy dose of audacity to establish a nonprofit organization to challenge the status quo in Washington, DC.

The *Oxford English Dictionary* defines audacity as "a willingness to take bold risks." On a humorous note, *Merriam Webster Dictionary* defines the same word as "a confident and daring quality that is often seen as shocking."

A *Harvard Business Review* piece on profitable audacity stresses the importance of audacious ideas, especially coming out of the 2008 financial crisis. The author, Vijay Govindarajan, states, "When there are severe resource constraints, it is more important than ever to maintain audacious goals."

Govindarajan goes on to argue that having a strategic vision that transcends present fiscal realities or constraints advances teams as well as corporations. "Because people are drawn to an audacious even unrealistic goal and they perform better when reaching for it. Performance is a function of expectations since we rarely exceed our expectations or outperform our ambition. It's where you set the bar."

"No one had a degree in renewable energy back then. You had to choose to be in the industry," said Mike to me when I spoke with him in the fall of 2021.

Since the beginning of his career, Mike always thought in terms of pursuing opportunities rather than taking on risks. To him, the risks come in the form of lessons when you fail and then assess why you took the risk and what went wrong.

"I really see the path forward as chasing opportunities and, as they unfold, pursing them in a somewhat calculated way," Mike notes. "We consciously did pursue a path, and the path we pursued was taking a risk in itself."

Mike attended Purdue University and received his Electrical and Electronic Engineering degree in 1973 after a distinguished career in the Navy. He went on to graduate from Harvard Business School with his MBA in 1975.

Mike's career afterwards spanned the conventional power business from doing beachhead studies at Booz Allen for the Carter White House on new energy technologies—like solar and wind—to planning nuclear power at General Electric (GE). In addition, his professional journey included being the chief marketer of coal fired power at Combustion Engineering in his mid-thirties to later on working with Aretê Ventures and United Power Systems. After almost two decades in the business, he decided to pivot to renewables in 1995.

At that time he made a conscious decision not to pursue coal-fired, nuclear, or gas-fired power. He had taken the years since the nineties to become an expert on conventional power generation but had hit a point where he didn't want to spend the rest of his career pursuing it.

"Year after year, I was disappointed in what I was working on," remembers Mike. "I was thrilled to be working at places like GE, but I was becoming more and more uncomfortable in dealing with things ranging from the brittleness of nuclear reactor vessels to the environmental dangers behind coal fire generation."

He sensed, from an engineering and finance perspective, "We in the US weren't quite getting it right, with engineers in the US being rushed to meet financial goals to make shareholders happy and meet sales margins."

As a thirty-two-year-old, he didn't have the position to challenge the status quo at his workplaces. With every advancement in his career, Mike felt the difference between the professional and commercial world: "Unlike the professional world I had known earlier, the commercial world had no rules of professionalism. It was all about getting a job done. Don't worry about morals and ethics, just get the job done."

He renewed his conversations about solar in the mid-nineties with those who had started solar ventures in the DC area, ranging from solar entrepreneur Neville Williams to a pioneering Hungarian Physicist named Peter Varadi. Varadi had started up a solar module manufacturing company named Solarex, which had been acquired by BP Amoco, or "BP" as we know it today.

Mike became immediately enamored by solar PV and the ability to apply the lesson of the Independent Power Producer (IPP) industry to the nascent solar industry in the US. The third party IPP model of having non-utility generators—that were not owned by public utilities—generating electricity for sale to national power grids and end users was a concept that had immense potential for the renewables sector. Mike sensed that.

Attracted by the "boy scout qualities" he saw in entrepreneurs advancing the wind, solar, and geothermal sectors, he liked "the culture and the dedication that people had to doing something good." Mike saw an industry built on qualities that included transparency, passion, intellectual curiosity, innovative spirit, and sincere longing to do right by the planet.

"I had found a home with my kind of people," recollects Mike. "Of course, I was taking a career risk, but I was getting out of situations that were to be far higher risk down the road in the traditional energy sector."

The conversations he found himself in were all about financing, and to Mike, the solar industry was stuck in the mode of focusing solely on cost and technology cost reduction. The questions of, "How are you going to finance it?" and "How are you going to create the conditions that would cause debt capital to support what you're doing?" were frequently reoccurring conversation starters in those early days.

Mike found the industry stuck in a rut. "We were stuck in the DOE funding mode, and I felt and saw the opportunity to unstick solar from its roots as a DOE or NASA contractor technology, bring it into the IPP model, and then bring it into the mainstream power generation industry."

He set about to establish the Solar Bank, which was built upon his strong personal motivation to use solar for the benefit of poor people primarily in the developing countries of India and South Africa. While setting up the bank, he did his research on who else was focused on solar deployment in the developing world.

Over the course of his first month, he reached out to his newfound network, beginning with Neville and Peter, and was introduced to Judy Siegel. Judy was pioneering the advancement of renewables in India as part of her role at Winrock International, which was an international nonprofit renowned for providing solutions to some of the world's most complex social, agricultural, and environmental challenges by empowering the disadvantaged, increasing economic opportunity, and sustaining natural resources.

Judy had already established an office in India with staff alongside John Bonda, a legend in his own right as the head of the European PV Industry Association. "All this happened in a period of weeks," recalls Mike, and before he knew it, he was in India and then South Africa meeting with local government stakeholders such as South African utility, Eskom.

While his attention remained on advancing solar loans to individuals with small businesses throughout India and South Africa, his thoughts kept returning to making an impact in the US. Given his international work, he found himself being invited to an international conference in Bonn in June of 2001.

Of the two hundred high-ranking global attendees from governments, corporations, academics and nonprofits, only two were Americans—Mike and Neville Williams. Supportive of having US participation, Hermann Scheer engaged Mike and Neville in a conversation of forming a US branch of his newly formed nonprofit, the World Council on Renewable Energy (WCRE). Neville had his hands full with starting up Standard Solar and a few other ventures he was working on, so Hermann challenged Mike. To everyone's surprise, Mike accepted.

"I thought creating an organization that united US renewables—and that would be part of something global—was exactly what we needed to get the US out of a thumping our own chest mind-set," recalls Mike.

While the US led in renewables R&D and technological development in the 1980s and 1990s, others—like Germany and Japan—had developed policies and financing mechanisms to catch up. Japan sparked its solar industry by launching its Sunshine Program in 1996, resulting in solar booming by 1999 with companies like Kyocera and Sharp making their entries into the solar sector.

Germany followed suit and in April of 2000, enacted The Renewable Energy Sources Act (EEG), which kicked off a series of German laws that originally provided a feed-in tariff (FIT) scheme to encourage the generation of renewable electricity. Mike didn't want to see US leadership in renewables marginalized and saw the need to unite the fragmented renewables industry. He was going to double down on a concept he had begun to engage in a few years earlier.

Acting on Hermann Scheer's ask, Mike gathered the movers and shakers he knew from setting up the Solar Bank for a series of breakfast meetings to discuss what would become ACORE—the American Council on Renewable Energy. He invited folks like Judy Siegel and Frank Tugwell from Winrock International; politicos Hank Habicht, former Deputy Administrator at the EPA, Dan Reicher, former Assistant Secretary of Energy at DOE, John Mullen from the USAID, Roger Ballentine from the Clinton White House, and Scott Sklar, a leading solar lobbyist; Michael Ware, president of Advance Capital Markets, Bill Holmberg, a biomass leader, Jim Woolsey, former head of the Central Intelligence Agency (CIA), as well as legal heavy hitters Mark Riedy and Roger Feldman and a few others.

The goal for ACORE was trail-blazing—to unite the various renewables technologies and revolutionize how the industry was going to scale up. The US renewables industry found itself fragmented as it approached the new millennium. Competing trade associations set up to represent each of the key technologies—primarily solar, wind, geothermal and hydropower—were fighting for their share of policies and fixated on developments on the Hill.

They considered the concept of ACORE audacious and to an extent moving in on their territory. "Incoherent and fragmented messaging is not a good strategy when we needed to make a compelling case for substantial policy change and a new energy economy," wrote Mike in a funding proposal to Michael Northrup at the Rockefeller Brothers Foundation in the early days of ACORE.

The original ACORE founders at that breakfast found themselves drawn to establishing a nonprofit social experiment that deviated from business as usual in Washington and set out to comprehensively connect all the dots through a strategic communications and policy platform.

They wanted an organization focused on three things: (1) making an air-tight economic and financial case for the deployment of renewables, specifically aimed at Wall Street; (2) encouraging the commercialization and operational professionalization of renewables technologies from Silicon Valley to the rust belts of the Great Lakes; and (3) filling in a policy gap by providing education versus lobbying to policy stakeholders on the Hill, convening policy-makers, financiers, and technology corporations as peers to the table and innovating policies that would grow out the industry.

ACORE was created to be a powerful solutions-oriented model to drive policy outcomes by being "for renewables and against nothing." For the first time, the renewable energy industry could speak with a broad but unified voice—an audacious goal that before ACORE's founding was unthinkable.

ACORE's "do it now and ask for forgiveness later" approach in Washington helped the organization push the envelope in setting up programs like the US Partnership on Renewable Energy Finance (US PREF) and the first ever all-renewable Renewable Energy Finance Forums (REFF) to connect financiers and law makers in understanding what each side needed to set up a stable investment environment. REFF-Wall Street specifically gave a national stage to selling the concept of renewables.

"We were selling Wall Street the idea that renewables are coming, and they're going to be just like gas-fired IPPs. We're going to use all the same legal structures... We just needed to educate the senior people in the banks and investment firms," remembers Mike on kicking off the first REFF-Wall Street at New York's Metropolitan Club in June of 2004.

At times, Mike felt he was a "refugee from the IPP industry that was adopted by renewables" and dedicated himself to educating bankers by challenging them on the applicability of PPAs, financing agreements, and non-recourse financing to advance renewables.

And that's what Mike did. ACORE and REFF provided vital market intelligence for how to grow the industry, featuring industry names such as Bloomberg New Energy Finance's Michael Liebreich, Susan Nickey of Acciona, Keith Martin of Norton Rose Fulbright, John Cavalier of Credit Suisse, Nancy Floyd of Nth Power, and Kevin Walsh of GE Energy Financial Service.

Reflecting on Vijay Govindarajan's *Harvard Business Review* piece and juxtaposing it against ACORE's mission, it was time to stop using economics and political discord on the environment as excuses to mute strategies surrounding the deployment of renewables. Indeed, they needed audacity more than ever. Govindarajan in his piece goes on to point out that, "Mediocre aspirations produce mediocre results," in his call to investors to take audacious corporate proposals more seriously.

Following in Govindarajan's footsteps, business performance coach Jim Canfield wrote an entertaining blog piece focusing on CEOs that don't present audacious plans or strategies to inspire their teams. In his piece, he quotes a famous scene from *Alice in Wonderland* where Alice encounters the Cheshire Cat. Alice asks the Cheshire Cat which way she should go. The cat asks where she's headed, and Alice responds that she doesn't know.

To Canfield, your responsibility as a leader is to set the direction for the organization. If you want your team to accomplish amazing things, he recommends setting big, audacious goals—or BAGs—with them.

"BAGs are about what's possible, not what's probable. They're the bull's-eye of a dartboard. Everyone still gets points if they miss the center but land on the board," notes Canfield. To him, the best BAGs are acceptable challenges that are both aspirational and quantifiable in that they can measure stakeholder satisfaction or performance overall for an organization or industry.

As leaders with audacity, messaging played an immense role in how Mike and the original ACORE founders framed dialogues around renewables. "We banned the word 'barrier' from the beginning," Mike says. ACORE's efforts in the early days were not only tied to creating a sustained institutional brand for the organization but also to build a unified brand identity for the renewable energy industry.

Mike grew ACORE while not taking a salary until a certain point. When he did take one, he took a low one to reinvest funding coming from membership and conferences into the growing venture. He believed in its mission and the impact it would have in giving the industry a firm foundation on which to scale. To him, it was worth the sacrifice.

ACORE sported a big-tent organizing model where all renewables technologies were invited under the mantra of being for renewable energy without being against anything else. This had a dual effect of attracting a variety of stakeholders to his oak tree, including those decidedly pro-renewables as well as those who would like to show support for renewables but couldn't be part of an organization that was against what they were already doing.

This subtle shift in mentality allowed me, as a marketer, to work with the ACORE team to convene a membership that grew to a peak of seven hundred organizational members by 2009. The membership composition was revolutionary for its time, bringing together all the renewable energy trade associations, government agencies, financiers, technology corporations, end users, utilities and power generators, and educational institutions united in the goal of advancing renewables deployment in the US.

Mike and the founders' vision was an organization to fill in the gaps to create an "all renewables" ethic, uniting the entire industry under one voice.

ACORE helped anchor that notion of one broad, coordinated voice by remaining nonpartisan and having the ability to move the dial on state and federal dialogue on renewables. They accomplished this through several efforts ranging from the industry's first comprehensive regional roundtables in 2005 to the Joint Outlook on Renewable Energy that was first published in 2007. All the while, the organization encouraged a long-term government approach to policies surrounding renewables deployment.

ACORE continues to succeed today as an oak tree for the renewable energy industry, having taken on the responsibility for empowering the industry's growth. According to the organization, renewables has grown to a share of 21 percent of the total electricity generation of the US grid, supporting 516,664 jobs in 2020.

In addition, the stakeholders driving the transition of the grid to renewables have evolved, with 69 percent of Fortune 100 companies having renewable energy as part of their energy mix in tackling greenhouse gas (GHG) emissions targets. The US is now ranked second in the world for attracting global renewables investment.

The mission of the organization remains true in developing breakthrough, nonpartisan solutions to achieve success for renewables in the US through high-profile events and strategic outreach to the business, financial, and policy communities. These solutions range from resource advocacy recommendations on ensuring low-cost reliability of renewables to providing stakeholders in the Southeast renewables power market options that maximize costs savings while ensuring emission reductions.

The ACORE Policy Forum continues as it has since 2004, and many other convening committees—such as US PREF and the Leadership Council—continue to bring together titans of industry alongside other leaders from the finance, technology, and power generating renewables sectors.

Reflecting on his shift to renewables, and later on in founding ACORE, Mike told me he definitely made a good decision transitioning from fossil fuels into renewables before renewables were a mainstream industry. He created his own path in having the audacity to create ACORE.

"I wasn't following others' leads," reflects Mike. "I intentionally decided I was going to be the leader, and others would follow. I was going to take personal responsibility for transitioning renewable energy into a model building on the

IPP industry method. I knew where we could take it. I wasn't taking it on with no idea. We had a very specific idea. I had 100 percent confidence that in working with others, we could do that."

CHAPTER 9

Social Entrepreneurship

Mindy Lubber

"Dozens of US businesses are ignoring the issue with business-as-usual responses that are putting their companies and their shareholders at risk," challenged Mindy Lubber speaking to stakeholders at the 2015 United Nations Climate Change Conference.

From the beginning of the early 2000s, Mindy felt that bringing the issue of climate change, along with both the risks and the opportunities, to the financial community was paramount. As a strategist and self-proclaimed formidable negotiator, she was passionate in spreading the message that economic development is intimately tied to social and environmental sustainability.

She made it her mission to convince investors, NGOs, and Fortune 500 CEOs to, "Adapt far-reaching positions on corporate practices to protect human rights and to minimize carbon emissions, water use and other environmental impacts."

As is common with renewables nonprofits, Mindy and I were introduced through mutual board members across organizations where we worked. The connection resulted in collaboration on a variety of high-impact finance and investor-related events such as the Renewable Energy Finance Forums (REFF-Wall Street and REFF-West) and several other leadership convenings. We worked together on international efforts being undertaken by the US renewable energy industry as well, specifically the Skoll Foundation and the Zayed Sustainability Prize.

Catching up on the ever-changing ESG and GRI market, Mindy and I recently reconnected again after a few years through our mutual acquaintance, Nancy Floyd, who had just retired from Nth Power.

Mindy began her career in environmentalism as an idealist fueled by a social entrepreneurial spirit. To her, climate change was more than just an environmental issue. "It's our generation's and our future's greatest social, environmental, public health, national security crisis threat. It's here and it's real," was her message at a TEDx Lake Geneva talk in 2014.

Comparing the US rebuilding after WWII or fighting for civil rights, Mindy believes each generation has its challenges, and climate change was our biggest challenge. She felt the call to challenge investors to invest the first "clean trillion" in changing investment patterns.

Unbeknownst to her in her early days with nonprofit CERES, Mindy and her team were going to be the forerunners for the Environmental, Social, Governance (ESG) movement of today.

The US Chamber of Commerce defines social entrepreneurship as, "The process by which individuals, start-ups and entrepreneurs develop and fund solutions that directly address social issues. A social entrepreneur, therefore, is a person who explores business opportunities that have a positive impact on their community, in society or the world."

One common misconception is that social entrepreneurs only have nonprofit 501(c)(3), (c)(4) or (c)(6) platforms to utilize in incorporating their call to action, but this is false. In addition to nonprofits, social entrepreneurship is alive and well in the for-profit corporate sector with one not having to look far to see examples.

The popular brand, TOMS, began a one-for-one social entrepreneurship model in 2006, in which for every pair of shoes sold to the public, a pair of shoes would be donated to children in need. Today, TOMS has given over ninety-five million shoes to people in need and has expanded their giving model to include access to safe drinking water, eye surgery, bullying prevention, and safe births.

Another brand that followed in TOMS' footsteps, which is familiar to those of us who wear prescription glasses, is Warby Parker. In 2019, the company launched the "Buy a Pair, Give a Pair" program, which to date has donated over eight million pairs of glasses. Going back further to early days of the company, they launched their "Pupils Project" in 2019, which worked with New York City and Baltimore local government agencies to provide free vision screenings, eye exams and glasses to school children who did not have reliable access to optometrists.

Aside from a purpose-driven idea, strategic vision, preparation, and research are integral to the social entrepreneur's initial attempt to set up a venture.

"If you feel called to make more of a difference in the world and want to make a living while doing it, you'll need a clear roadmap for your journey," comments C.J. Hayden, a San Francisco-based social entrepreneurship coach and founding board member of the San Francisco Bay Area Chapter of the Social Enterprise Alliance.

Speaking on *The Thriving Women in Business Giving Community* podcast, she adds, "A specific set of directions to get your business from a flash of inspiration to off the ground is required."

CERES, which stands for "Coalition for Environmentally Responsible Economies," was formed in 1989 as a response to the Exxon Valdez oil spill, when a group of forward-looking socially responsible investors and environmentalists led by pioneer investor Joan Bavaria came together in Boston.

Their vision was to change the way businesses operate and, "redefine the role and responsibility of companies as stewards of the environment and agents of economic and social change." They aimed to create their own oak tree that would convince businesses to consider their impacts on the environment, employees, and communities and actually do something about it.

But in doing so, Mindy understood that to move the marker, the industry needed data to benchmark progress. "The big thing that followed was that we realized nobody had data or good information, nor a uniform approach to gather the data."

In 1997, CERES crafted the first ever code of environmental conduct for companies and launched the global standard for corporate sustainability reporting, known as Global Reporting Initiative (GRI), which is used by over ten thousand companies worldwide.

This had placed the organization on the radars of many, including Mindy, and gave birth to the corporate sustainable responsibility (CSR) movement that initially focused corporate leadership on recycling, energy efficiency, and waste streams.

As CERES ramped up, Mindy was just completing her role as a Regional Administrator at the US Environmental Protection Agency under former President Bill Clinton from 1999 to 2001. She had come into the administration after a long history of working on environmental finance issues in Massachusetts, first founding Green Century Capital Management and then serving as the director of the Massachusetts Public Interest Research Group (MASSPIRG).

All throughout her career, she felt a call to make a case for the integration of sustainability into the capital markets. She wanted to highlight the important need for corporations to merge their sustainability and finance reports in order to approach business in a whole new way based on the triple bottom line of profit, people, and planet.

Her suggestion to introduce people and planet as equally important priorities to a profit was unchartered territory in corporate boardrooms, but her next career move would give her the platform needed to tap into her skill sets as a social entrepreneur and figure out a way.

Coming on board to lead CERES in 2003, Mindy and her team—at the invitation of Tim Wurth of the UN Foundation—hosted their first Investor Summit on Climate Risk at the United Nations. Their goal was to impact how major players were approaching their capital markets investments around the energy sector by considering the yet to be quantified economic consequences for continued investments in fossil fuels.

They invited CEOs from all the major banks and investment firms and got pushback from several of them. Instead of top-level executives, these firms offered junior team members covering the nascent environmental finance sector.

Mindy recollected that more insult to injury came during conversations with investors who would decline speaking on the topic and instead would point Mindy and her team toward philanthropic foundations to get funding to provide research on this "cute idea."

The financial sector was not recognizing what Mindy saw in front of them. To her, they were about to have to transform the base of the US economy to a different one that would require the revamping of the US's entire energy and transportation system.

According to a foreword in CERES' Corporate Governance and Climate Change Report in 2003, the evidence was increasingly compelling:

> Companies' performance on social and environmental issues does affect their competitiveness, profitability, and share price performance. And climate change, arguably one of the world's most pressing issues, exemplifies the challenge better than most. A company's response to threats and opportunities of climate change— or their lack of response—can have a material bearing on shareholder value.

Mindy was determined to show the impact sustainability would have on the corporate sector's triple bottom line. "It was not just about the balance sheet for the next six months or three years," she adds. "It was about everything else they do, and the language was different. The sophistication of the arguments was different. The economic impact studies were different. The data was different, but that was the premise."

Strategically building a power base of asset owners on her board who managed the biggest institutional investments, she assembled a constituency of significant asset owners that those money managers had to listen to. These asset managers represented several leaders with diverse backgrounds in sustainable investing, business, research, policy, law, and advocacy.

Partnering with entities such as the Investor Responsibility Research Center brought additional big names into the fold, including Ford Motor Company, American Electric Power (AEP), ChevronTexaco, DuPont, and International Paper Company. The united voice of these asset owners could not be ignored.

The message to money managers was simple: "We see climate change as a financial imperative, as a material risk, as a governance issue, and as something that will drive the future of our economy, and we need to pay attention to climate risk and the opportunities around it."

CERES' message in turn was clear to the financial and corporate sector: "You have to be more transparent with investment and cut financing of the fossil fuels."

They called on the "big six" institutional investors on Wall Street, which included BNY Mellon, Citigroup, Goldman Sachs, J.P. Morgan, PIMPCO, and Vanguard among others to back up their climate action plans with more than putting money up to finance renewables.

Mindy knew what these large investors were thinking: "They had this mind-set that if they didn't finance a fossil fuel project, their clients would go down the block and somebody else would finance it, and they'd lose the client."

It was clear to Mindy that financial objectives had to be married with a societal call to action that had more teeth to it than CSR standards already out there. The timing could not be better with the entrance of environmental, social, and governance—or as we call it ESG—accountability.

Tracing the origins of ESG investing takes one back to the 1960s when investors began to exclude stocks or entire industries from their portfolios based on business activities such as tobacco production or involvement in the South African apartheid regime.

The term ESG was first coined in 2004 in a landmark International Finance Corporation published study entitled "Who Cares Wins." ESG received a boost during that time from Socially Responsible Investment (SRI) movement that took hold in the middle 2000s, which in turn was based on the original CSR movement begun in the 1990s.

With ESG factoring into corporate decision-making, CERES finally had the attention and traction it needed. It went on to launch a variety of initiatives including an audacious initiative in 2014 called the Clean Trillion Challenge. The International Energy Agency (IEA) has estimated that a thirty-six-trillion-dollar increase in investment in renewable energy by 2050 can keep us below the two-degree threshold.

CERES launched the Clean Trillion challenge calling on businesses, investors, and policy-makers to take measures to close the investment gap by one trillion a year for the next thirty-six years to meet that target. If Fortune 100 invested 0.1 percent of their profits today, Mindy affirms, "We'd already be halfway to the trillion-dollar goal."

As a social entrepreneur, Mindy believes corporate leadership in tandem with customer demand and "the power of the purse" gets us to this goal. "We've gone from words to deeds and from roles to plans," observes Mindy. She added that we're starting to see tangible results in moving corporate and financial mind-sets in how they approach ESG and specifically sustainability.

"When you'd in the past go to talk to the sustainability lead at these large companies, they were at best middle management," Mindy recalls. *Contrasting the vast difference today, she notes, "Now we're finding sustainability people in the C-suite and the Training Program with the University of Berkeley Law School is not able to satisfy all the numbers of corporate board members coming through wanting to understand why climate and sustainability is a governance matter for the world's largest companies."*

At first, social entrepreneurs like Mindy who fought on issues surrounding the climate crisis struggled to compete in getting their missions heard in the overall US business environment. Their ability to be agile thinkers in using their knowledge of the energy and corporate sectors allowed them to create a social movement. That movement, when united with things like ESG or long-term risk alleviation, sparked revolutions in how businesses viewed renewables and sustainability.

Social entrepreneurs "Put purpose in front of profit," explains Mario Calderini, Professor of social innovation at MIP Politecnico di Milano in *BusinessBecause*. "And the movement is growing—30 percent of all social enterprises were founded in the last three years." Just like for-profit entrepreneurs, social entrepreneurs aim to scale, maximize profits, and increase market share. But they do so not with customer-facing products or services. They do so with their ideas.

ESG today has brought the objectives of social enterprises and corporates ever closer, redirecting impact investments into companies that not only are doing well on profit margins but also equally emphasizing a long-term sustainability perspective alongside a culture of purpose.

According to Dr. Ken Read, investment in companies ranking highly for ESG stands at twelve trillion dollars in the US in 2021, a 38 percent increase since 2016. This subset of a new approach to capitalism, which he calls "new capitalism," empowers businesses to be designed to add value to society in multiple ways.

These businesses in turn work for the benefit of all stakeholders, ranging from customers and employees to shareholders and communities. Business values can be tied to a host of issues, ranging from sustainability to equality, transparency, environmental impact, and social impact.

This combination of purpose and profit continues to revolutionize the approach many corporations take today to conduct business. In turn, their approach bolsters the causes and the impact of social entrepreneurs in the market.

Désirée van Gorp, Professor of International Business at Nyenrode Business University in her recent interview with *BusinessBecause* agrees: "If successful, ESG investment could finally allow social entrepreneurship to have a large-scale impact and move from supporting small communities to tackling society's most pressing issues. That entrepreneurial spirit, if it is done well across different areas, has the potential to do so much and to do it is so effectively. I think it is the only way to find solutions to global challenges."

In comparing her role in government versus her role today as a social entrepreneur at CERES, Mindy tells me that by going the nonprofit route, "The burden was so much higher. You don't have any real power, and instead you've got to build it. You've got to design it. You've got to do so with convictions and fresh ideas."

Despite this, she wouldn't have changed her career path at all. "At the end of the day, it's about focusing on the most effective path to get to where you need to go. Sometimes it's going to be tough, regardless of where you are, but don't let go of your volition to continue forward because this cause we're fighting for is really important."

CHAPTER 10

Adventurous

John Cavalier

"The entrepreneur has driven the success in the renewable energy industry. Let's protect the entrepreneur as we go forward. We are at a wonderful inflection point as an industry. Let's seize the day," began John Cavalier's speech at the RETECH 2009 conference.

For renewables investors or anyone who has spent time on Wall Street, John is no stranger. Known for the principal roles he played in the Initial Public Offerings (IPO) of SunPower, Suntech, REC, Cosan, First Solar, and Iberdrola Renewables, he stands among the giants we all associate with renewables financing on Wall Street.

While at the conference, John went on to further reiterate his belief that, "Strong businesses with good management, sound business plans, and viable technologies will find financing. However, when the cost of capital escalates and fluctuates, capital does freeze. Thus, to initiate the flow of capital, we must have stability in the financial sector.

"With the Stimulus Bill, the administration has opened the door, but the industry must take advantage of the opportunity. Consumers need to be educated if they are going to be supportive of renewable energy and climate policies."

I had first encountered John at ACORE's REFF-Wall Street when introduced by an old friend at Credit Suisse, Nick Sangermano. John, in my mind, was a giant of giants, who was eloquent in how he described the financial challenges surrounding investments in the industry, but he often struck me as being adventurous and scaling the heights of Wall Street with the purpose of showing that renewables not only were financeable but profitable.

But like all risk-takers in renewables, his beginnings stemmed from humble origins tied to a courage to embark on a journey where few have gone before in the financial sector. The adventurous spirit John Cavalier embodies in approaching renewables finance began when he was growing up in Spanish Harlem.

John was a troublemaker, notorious for getting into at least one fight a day. After losing his father when he was three years old, his mother moved them in with his grandparents on 172nd Street between St. Nicholas and Audubon Streets.

His mother and teachers saw potential when testing scores uncovered a high aptitude for learning. Things started to turn around when he took an entrance exam and got a scholarship to attend the prestigious Regis High School. Five years later, he found himself attending West Point Academy in Hudson Valley, New York.

John majored in environmental science and found himself in a small cohort made up of two other cadets, who shared his interest in the environment. "The technology boom was just really taking off and that's where I really connected," recollects John when he spoke to me in January of 2022. His adventure into renewables was just beginning.

According to American psychologist Matt Walker, "adventure" is defined by uncertain outcome. In interviewing numerous adventure seekers, he found that in every case the most significant moments of these adventurer's lives—ranging from the most important decisions and the most meaningful choices—are characterized, in part, by uncertainty.

Walker claims that without uncertainty we have a safe, contained, and predictable experience, albeit one without adventure. Instead of expecting the future to deliver something specific, he encourages adventurers to focus on what they will do to create the experience and outcome they desire. Walker claims, "Your actions and intentions are within your control," ultimately impacting the adventure—and the ultimate achievement—you're aiming to target.

After John graduated West Point in the 1970s, he spent nine years in the US Army, where he was an officer in the Signal Corps and later the Judge Advocate General Corps. During this time, he was awarded the Meritorious Service Medal for his role as the first Senior Defense Counsel for the Northeastern US.

John's quest for knowledge was not yet over, and he found himself getting a JD from the University of Illinois School of Law and later an MBA from Harvard Business School. During this time he kept up with clean technology and sensed a need to get more serious on how to finance entrepreneurs with technologies that were going to redefine the next evolution of the renewables industry.

That opportunity came in January 1990 when John started his finance career at Donaldson, Lufkin and Jenrette (DLJ), and Drexel Burnham Lambert. Eight years into his job at DLJ, he got approval in 1998 to start an Energy Technology Group, and they started to invest in early-stage small solar, superconductor technologies, fuel cells, and flywheels. One of their notable investments was AstroPower out of Newark, Delaware. John's adventure in the financial sector moving forward would be tied to the IPOs he successfully ran.

Investopedia defines an IPO as the process of offering shares of a private corporation to the public in a new stock issuance. An IPO allows a company to raise capital from public investors. The transition from a private to a public company can be an important time for private investors to fully realize gains from their investment as it typically includes a share premium for current private investors. Meanwhile, it also allows public investors to participate in the offering.

Founded in 1983, AstroPower was one of the first fully integrated solar companies in the market during the eighties and nineties, manufacturing and marketing solar cells, modules, and systems for generating electric power from sunlight. With John's help, the company became a publicly owned company with the launch of its IPO, traded on the NASDAQ. He had found his success in launching renewables companies through the IPO process.

But in order for John to believe in launching an IPO with a company like AstroPower, he had to know that the team believed in their technology and had a passion for adventure. "The AstroPower team was a brilliant group of people who exhibited a real passion around their technology," notes John. "In order to invest and believe in a team, you really had to like them."

During his time leading the Group at DLJ, which later got acquired in 2000 by Credit Suisse First Boston—now just Credit Suisse—John wasn't afraid of mistakes and failures. Rather, they fueled his ambitions to seek out additional companies that would leave their mark on the industry.

Concurrently, he started to follow renewables conversations and trends in Germany, which was beginning to advance its renewable energy ambitions. Of the notable figures leading the charge was a rising Green Party parliamentarian named Hermann Scheer.

Dubbed the "George Washington" of renewables by colleague and friend Mike Eckhart, Scheer, along with fellow Bundestag member Hans Josef Fell, are credited with pulling together Germany's feed-in tariff (FIT) policy in the early nineties, the first policy of its kind for the renewables industry globally.

Scheer's leadership resulted in Germany being the first country to amass 60 percent of the wind farms and 70 percent of the solar farms in the world as part of their energy mix in the late 1990s and early 2000s. These policies sparked a flurry of corporate entities focusing on renewables technologies—such as Conergy, Enercon, Nordex, REpower Systems, Siemens, and Fuhrländer—in Germany, dominating the renewable energy landscape.

Scheer ultimately succeeded in Germany's adoption of renewables and furthermore helped to launch the International Renewable Energy Agency (IRENA), which would raise global attention and commitments to renewables deployment.

These results intrigued John, and he wanted to replicate this success in the US. During this time, John's team was trying to understand which technologies were available to market. They investigated which ones had sound technology and focused on both increasing the efficiency of energy production while avoiding the creation of pollution.

"We saw early-stage movers making difficult capital decisions when entering the market," recollects John. To him, these early movers' experience entering the market was impacted by the fact that, "No one had a record of success or failure yet." What came next only made the landscape muddier.

The 2002 bubble burst came without warning, and as John remembers, "It was devastating because it took down an extraordinarily resilient energy technology sector." This setback did not stop him and his team. Despite the burst, the renewables industry had momentum and the European market, though not insulated from the ramifications, continued to grow.

As the industry grew post-2002, so did John's career, and the best adventures were yet to come. Weathering the aftermath of the burst, he focused his time on other investments that were growing in the energy sector.

He climbed the ranks of Credit Suisse to become chairman of the Global Energy Group, overseeing 164 bankers that focused on everything from investments in metals and mining to infrastructure and oil and gas power utilities in the US, Canada, and Latin America. But his ambitions and adventures in scaling the barrier walls of renewables finance were just beginning.

In 2005 he decided to approach the two co-heads of investment banking at Credit Suisse with a challenge to set up the first ever Renewable Energy Banking Group—a feat not yet accomplished among the investment banking giants in Manhattan.

He knew the question of how individual groups would receive credit for profitable investments would drive the decision to approve establishment of the new Group, so he asked for an exception that, again, was unheard of in the finance world.

"I need large groups within the firm to be able to share credit for every renewables deal we, as Credit Suisse, do," was the beginning of his ask.

With no deals in renewables going on at Credit Suisse, he met some pushback, but this equitable way of receiving full credit for a deal—no matter if a deal originated within this newly conceptualized Group or outside of it—won good will among the four investment banking groups at Credit Suisse that would compose this new entity: the Technology Group, Industrial Group, Energy Group, and general Infrastructure Group.

This ability to crosspollinate efforts across these groups resulted in John also getting approval to reward three-to-four-times credit for a deal originated among these groups. This adventurous new territory for John and Credit Suisse was tied to a gamble on the potential risk payoff for this new cross-cutting group within the Global Utility, Infrastructure and Renewables Group.

Leveraging the new Group, and after learning from their IPO experience with AstroPower, he embarked on his next IPO in 2005, this time with SunPower Corporation, a subsidiary of Cypress Semiconductor Corp, with joint book-running management shared with Lehman Brothers. For John, his proudest achievements were tied to the stories and adventures around launching some of the most notable IPOs in the market.

SunPower was a success story in itself that mirrored the beginnings of the solar energy industry in the US. It was founded in 1988 by Richard Swanson, who had been a classmate of Cypress Semiconductor Chief Executive T.J. Rodgers. After years of refining SunPower's cell technology and encouraging the adoption of solar power in Silicon Valley, Swanson was two weeks from laying off half his company when he caught the ear of Rodgers in 2001.

According to a *Mercury News* article, Rodgers wrote a $750,000 personal check to Swanson on the spot and then approached his board of directors to make an investment. The board was wary, but Rodgers pitched it as part of Cypress's diversification strategy that could reduce its dependence on the volatile memory chip market. Cypress went on to buy a controlling stake in SunPower in 2002 and the rest is history.

A major reason for John taking on the SunPower's IPO was based on his relationship with SunPower's visionary CEO, Tom Werner. John found that an integral part of his success was tied to the entrepreneurial leaders he was encountering during his due diligence conversations in the growing renewables industry.

Having taken the helm as CEO in 2003, Tom had thrived in the initial days of Silicon Valley's embrace of the solar energy industry. Before joining SunPower, he was CEO at Silicon Light Machines, Inc., an optical solutions subsidiary of Cypress Semiconductor Corporation.

Prior to that he held the role of VP and general manager of the Business Connectivity Group of 3Com Corp. Tom was passionate about investing in companies that would lead to a better, cleaner, and more equitable future while reversing the impacts of climate change.

He believed in SunPower and that the company could change the way the world was powered. Tom saw the day coming when solar was going to be competitive against mainstream energy. Solar was no longer a lab project to him. "High-tech solar power is now commercial and competitive with the highest power rates in certain parts of the US," he said in an interview with *Forbes* following the IPO. Tom continues to play an integral role in SunPower, having stepped down as CEO in 2021, and remains as Advisor to the CEO.

John used SunPower's $150 million IPO to test the waters in identifying winning entrepreneurs and technologies with the right business models. The adventures many of these CEOs had taken in pulling together their start-ups fascinated John.

"They weren't afraid of the unknown, and they thought they knew what should happen," reflects John. "They weren't afraid that no one else had gone there before them and so they had the courage and excitement needed for their journey."

His SunPower adventure led to John meeting another individual who he to this day considers an integral part of his adventure—Dr. Shi Zhengrong the founder of Suntech Power.

Suntech Power was founded in 2001 in Wuxi, China, as a powerhouse solar module manufacturer. John saw a visionary with a persistent work ethic in Dr. Shi, who at the time was still teaching part time about solar power at the University of South Wales while he was setting up the company.

He met Dr. Shi at the Beijing International Renewable Energy Conference (BIREC) in 2005, and the two decided it was time for the company to take a next step. Later that year, Suntech's $450-million IPO on the New York Stock Exchange was a first for a privately owned Chinese company. John had accomplished what many thought impossible.

Aside from making Dr. Shi the richest man in China in 2005, he had succeeded in launching yet another solar company into the public offering, even though many still did not understand the true trajectory for renewables. Dr. Shi went on to be named a *TIME* magazine "Hero of the Environment" and China's "Green Person of the Year" and continued to see the company through its ups and downs until its demise in 2013.

According to John, the reason why investors were willing to oversubscribe to these IPOs was due to the successes they saw in Germany. "SunPower and Suntech revealed a phenomenal pent-up demand to put money to work behind a new technology that would be clean," he states.

John saw the opportunity to tap further into this demand and moved forward with another successful IPO in May of 2006, this time with a Norwegian entity named Renewable Energy Corporation (REC). REC, which produced vital crystalline silicon wafers needed for the growing solar industry, was part of John's strategy of focusing on investments in companies located in the beginnings of the solar supply chain.

He believed these companies were set to grow exponentially once the industry recovered. John's gamble paid off. The IPO was a success and the company was the largest wholly privately owned company in Norway since the privatization of Statoil in 2001.

Of all the IPOs that John helped to advance in the solar industry, he felt two specific IPOs showcased the immense growth and strides undertaken by the renewables sector in the early 2000s. Those significant IPOs were First Solar's IPO in November of 2006 and EnerNOC's IPO in May of 2007. Both companies were led by two more visionaries that helped define the industry.

Headed by Mike Ahearn, First Solar defined a space in the market, seeking to enable solar electricity to achieve price parity with conventional energy while reducing fossil fuel dependence, greenhouse gas emissions, and peak time grid constraints. He had come from a legal and private equity background, first practicing law at Gallagher and Kennedy, and later partnering in an equity investment firm JWMA (formerly True North Venture Partners, LP), focused on investments in early-stage companies in the energy, water, agriculture, and waste sectors.

Mike had received initial funding from a venture fund organized by Walmart heir and philanthropist John T. Walton. After John's tragic death in an experimental, ultralight aircraft crash, investments ceased from the fund, and Michael was left to search for capital to grow First Solar.

John and Michael joined forces, and through the successful IPO launch, they raised the capital needed to help bolster First Solar's growth. Mike is still at First Solar, serving as chairman of the board today.

Led by Tim Healy, the founder alongside David Brewster, EnerNOC had made a name for itself as a maker of energy management systems for utilities and enterprises since its creation in 2001. EnerNOC did something revolutionary and used a network operations center to remotely manage electricity consumption across a network of end-use customer sites, making electric capacity and energy available to grid operators and utilities on demand.

Tim was known as an entrepreneur with his founding role at EnerNOC eclipsed by his cofounding role in setting up Student Advantage, a college student discount program that most of us recognize as the logo at bookstores and retailers that saved us money on college supplies.

He and his team were obsessed with changing the way the world uses energy through demand response, and their business model had picked up several clients in Massachusetts. It was now time to expand EnerNOC outside of Massachusetts and the company needed to raise capital. Tim had turned to John for advice on launching an IPO.

John's key advice was, "Just like a real baby, you have one shot to deliver this company publicly. Be sure you do it the right way." He added that when you price your shares be sure not to overprice, because if you do, "You're not going to get any forward momentum. You're only going to be able to slip back."

Though Credit Suisse ultimately didn't cover the IPO, it went public with an $85 million IPO. After sixteen years of success and with Enel's acquisition of the company, Tim moved on and continues to work on entrepreneurial ventures.

During all his transitions and adventures from one IPO to another, John stated the importance of being relevant in the market. "It's hard to give up continuing your pursuits because you want to be relevant." He notes that, "Relevance gets derived from the relationships you've got with people who are active and doing things."

John's adventures in the industry had built an oak tree that attracted, gathered, and inspired a whole generation of bankers and investors to the US renewable energy market. The relationships he made, coupled with his honest approach to banking, made him a permanent name on Wall Street.

His relevance in the renewable financial markets continued with the launching a few more IPOs, most notably Iberdrola's lucrative $5.5 billion IPO in Madrid Stock Exchange in December of 2007. That IPO, which was a bear in itself to conduct, was the second largest IPO for the financial sector in 2007 according to *The New York Times.*

With Credit Suisse's blessing, he went on in 2007 to cofound a private equity firm named Hudson Clean Energy Partners to invest in the dynamic and fast-growing renewable energy market.

Hudson's fund made a debut three years later in 2010 with commitments of $1.02 billion. The firm went on to make privately negotiated investments in infrastructure platforms focused on the development, construction, and operation of power plants.

In addition, the fund made investments in the high-growth, asset-based, capital intensive segments of value chain companies that have manufacturing or servicing businesses. The solar industry was growing quickly out of its infancy, but as with all things, the adventure in pushing out one success after another was going to abruptly pause.

Growth of the solar industry was pegged to a vital commodity—silicon. Silicon prices were low during most of the 2000s, averaging around thirty dollars per kilogram. In 2010 into 2011, the prices suddenly skyrocketed to around one hundred dollars per kilogram and then up to four hundred and fifty dollars at its peak. Investors were just seeing upside and pumping money into companies like REC. And then in the height of the frenzy, silicon prices crashed.

The crash was due to a number of factors ironically tied to better manufacturing technology coupled with economy of scale and the entry of fully integrated Chinese manufacturers into the market. According to the blog, *Down to Earth*, the drop in price was also related to the economic crisis back in 2008 that suddenly created a heavy oversupply as European and US growth of PV demand slackened.

This oversupply squeezed the profit margins of Crystalline PV manufacturers as they competed for a smaller market pie. For the industry, this time period between 2008 and 2011 was a dark time with bankruptcies of Solyndra and US Crystalline manufacturers, Evergreen Solar, and SpectraWatt.

"When an industry is experiencing that kind of growth from the very infancy of that industry, things are getting commoditized and then price is a commodity," John tells me. "When that commodity is scarce, and then plentiful, you can't make a smart investment decision. It's an impossibility."

The shock of silicon pricing combined with his forty years of experience in trying to get solar thermal off the ground taught John what he referred to as an "expensive education" that he said resulted in, "scars on my back," but were all part of the adventure of being the first one out there.

In looking back, not only do we now have the benefit of seeing all that capital mobilized by the "Big Six" on Wall Street, but we can also now mobilize more specialized capital from pension funds and other institutional investors with the benefit of understanding how to manage commodity risk across the entire renewable energy spectrum.

The ability to see how that initial capital was dispensed into the market, the way the market reacted to it, and how sophisticated we got with the financing remain foundational cornerstones to those of us structuring renewable energy deals today.

John never saw negative outcomes as failures during his adventures in renewables investing. Reflecting on what he would have told himself, knowing what he knows today, he reflects, "I've made a number of mistakes but wouldn't change that. If we didn't try, we wouldn't have known if the technology was viable. And the fact that as humans, we were looking for alternative ways to create energy that might not adversely impact the planet. Doing just that was enough."

In much the same way, T.S. Eliot in his preface to Harry Crosby's collection of poems, *Transit of Venus*, stated, "Only those who risk going too far can possibly find out how far they can go." The adventurer in John continued forward.

In 2019, he moved on to Braemar Energy Ventures to lead as a venture partner. True to John's adventurous ways, in 2020 he embarked into the world of special purpose acquisition companies or SPACs, a phenomenon that has taken the financial markets by storm.

"A catharsis comes when you dust yourself off, say to yourself, 'Okay, let's do this again, but let's do it differently,'" exults John. Because of the experiences of Hermann Scheer and the visionary CEOs behind the IPOs John launched, we have learned the "ropes of the industry" in how to get up, dust ourselves off, and continue, despite our failures.

Much like a scenario found in a "Choose Your Own Adventure" book from the 1990s, one has to make calculated choices and persist as a risk-taker but do so with a sense of adventure. In order to make the right choices to get to your goal, you need to learn lessons along the way and use them to forge your adventure.

"One thing is sure," John ends when reflecting on his adventures in renewables finance. "If you don't know the history of renewables, you won't know the future."

CHAPTER 11

Foresight

Michael Liebreich

"Starting New Energy Finance for me was like rappelling into the abyss. Here I was—a Cambridge engineer, Harvard Business School, Baker scholar, first class, everything's great, but I was high and dry, unemployed," harks Michael to his early days in renewables.

Michael Liebreich's reference to a scene from a book by Joe Simpson called *Touching the Void*, in many ways parallels the situation he found himself in before making the decision to transition into renewables after a career in the dot-com era.

His foresight into the potential renewables held for the future of the energy sector led him to create his greatest venture to date, New Energy Finance (NEF), now known to many in the renewables industry as Bloomberg New Energy Finance (BNEF).

In the book, the main protagonist is a climber who is set to embark on a great saga of climbing a challenging mountain in the Andes with a buddy. Halfway on the trip up, he ends up falling into a crevasse. After hours of trying to save him and lowering him further into the crevasse to cling to a ledge, his climbing buddy makes the decision to cut the rope and leave him behind.

Unable to get out of the crevasse due to a broken leg, the protagonist has to assess his options: either stay and die, while clinging to the hope of an eventual rescue that will not come or take the situation into his hands by lowering himself into the abyss below with a fifty-fifty percent chance of death while seeking a potential way out. The climber lowers himself into the abyss and eventually finds the way out.

I met Michael at the same time I met John Cavalier at one of ACORE's first Renewable Energy Finance Forums (REFF-Wall Street) held at the Waldorf Astoria in 2005. Shortly after, I worked with Michael to bring a short-lived ACORE-NEF digital newsletter to life with initial aspirations that it would help launch NEF out to the market across North America.

In 2006, I had the privilege of helping Michael recruit and interview his first US hire, Ethan Zindler, who to this day remains a friend and the Head of Americas for BNEF. I was fortunate to have Ethan working down the hall from me at ACORE when NEF had sublet space, and I miss the days when I had a random industry question and could walk over to Ethan to garner his advice.

Michael remained a true friend to me and the ACORE team as he built his empire, and the lasting trait that always came to mind when thinking of Michael was his foresight. He has this insightful instinct about the renewable energy financial markets, along with an ability to model key future trends and accurately forecast technology pricing. Whenever he comes out with market insights and recommendations, every banker following renewables—from Wall Street to Canary Wharf to Shanghai—cues up to read his next prediction.

The *Collins English Dictionary* defines foresight as, "The ability to see what is likely to happen in the future and to take appropriate action." What differentiates vision from foresight is the ability to not only envision what the future holds, but to take definitive action to capitalize on that vision. With that vision, one has a choice to either keep the status quo and capitalize on any pre-envisioned shifts that occur in the future or work to positively redefine it in the hopes of creating a better future or more positive outcome.

In a related *Harvard Business Review* piece entitled "How to Do Strategic Planning Like a Futurist," author Amy Webb states that to have foresight that is truly impactful in business, one-, three- or five-year strategic plans, though useful for addressing short-term operational goals, do not provide the foresight and inspiration needed to plot a sustainable vision or set of goals.

In her thesis, she claims that futurists specifically think about time differently and not in terms of "short and long term" simultaneously. Instead, futurists like Michael use a time cone that measures data points associated with certainty, scenario planning, and action charts.

Looking back, the abyss started out as a dark one for Michael, and despite his foresight allowing him to see a light at the end of the tunnel, he couldn't predict the journey in front of him. He had a successful start with a career at McKinsey, a trusted global advisor and counselor to many of the world's most influential businesses and institutions, which lasted close to five years followed by a venture starting up the first travel internet site in the UK in 1997.

In 1999, he was brought in by Bernard Arnault, chairman and CEO of Louis Vuitton Moët Hennessy (LVMH) to build up a portfolio of internet-related businesses dubbed Europe@Web on behalf of Groupe Arnault. Michael built up a professional team and ran the group's tech investments for, "As long as it took for the dot-com bubble to completely blow up," says Michael.

Michael had options in the Arnault internet venture that were at one point valued at thirty million dollars. From the beginning, he had the intent of reinvesting his earnings in a new venture, though he admits, "I didn't for an instant believe it was worth thirty million." He thought that with Arnault's backing, the assets could always be bought up if anything happened.

The dot-com bust came swiftly in the early 2000s, and Michael lost 99 percent of what he thought he would get from the portfolio while also finding himself unemployed. Thankfully, he had some resources at his disposal from his previous travel venture.

Having descended into a destabilizing period in his life where he thought he was "unemployable," Michael decided to put his athletic skills in mountain climbing as well as being a former skier for the British Olympic Team to the test again as he journeyed to clear his head in Bolivia and Brazil. All throughout his trip, he saw the impact of pollution at high altitudes impacting climate seasonality and mitigating snowpack.

Indeed, studies coming out years after Michael's expedition corroborated what Michael had seen. According to a recent University of Syracuse report, pollution particulates, which are small, light-absorbing impurities that are produced from the incomplete combustion of carbon-based fuel, are starting to have a major impact in snowpack high up in mountain ranges around the world. When deposited on the snowpack, these particles reduce snow albedo and accelerate melt.

What Michael experienced, coupled with what he saw as aging energy infrastructure supporting a fragile economy, demanded that humans adapt to a changing environment. He needed to figure out how to connect this newfound passion to his next business plan. His foresight yet again would be an invaluable trait in powering his ability to adapt to his new venture.

As with any entrepreneur who stumbles upon an opportunity in the market, Michael arrived back in London recharged with a few business ideas in his head. He started by assessing his situation.

Michael had his savings and a rolodex of unemployed colleagues who were fatalities of the dot-com bust. Diving into his thoughts from his trip to South America, he engaged in forward-looking market research and conversations in the energy sector about peak oil, local pollution, and up-and-coming technologies in what was being termed in 2003 as the clean energy sector.

Michael foresaw a plethora of new renewable technologies being supported by various early market entrants to the market across Europe and the US, including a few venture capital firms, some early adopter technology giants, several European wind companies, and a handful of solar technology companies.

Applying his Cambridge education in aeronautics and fluid dynamics, hydrogen and fuel cell technology caught his eye as a safe sector to rebuild his career. After further research into the hydrogen market, its supply chain, and the overall state of the technology, he determined the technology was not yet there and decided to broaden his scope to assess the clean energy technologies he felt were going to take off—wind and solar.

From the get-go, Michael acknowledges, "I wasn't really a renewables guy. I was the information provider."

All throughout his assessment of this new energy sector, he realized there was not widely reliable and updated data in the market. Aside from some notable reports from energy giants Shell and BP, no industry-wide database of the technologies existed with financiers and no forward-looking business projections could spur this nascent new sector.

Attending initial "industry-first" conferences in 2004, such as the first ever REFF-Wall Street in New York, he quickly realized the broader clean energy approach was the way to go with no one really talking about hydrogen.

Michael's concept of a database was garnering traction with potential customers onsite. His vision for the database included sophisticated datasets and baseline energy models that would track clean energy production and pricing across the global industry.

Many in the investment sector stated they wished they were doing what he was doing in assembling these industry statistics, but no one had budget for such a database. Better put, no one had the compelling argument in place for how this database would positively impact their investment strategies.

The market didn't necessarily welcome him with open arms, and he started receiving a multitude of critiques and recommendations on how to make the data better and more useful. These included ideas on complementing the database with published reports as well as commentary, insights into technology trends, reactions from industry on those trends, and long-term forecasts every year.

Michael knew he had to adapt to this new industry and assess what his new clients' challenges and needs were.

"And instead of saying you're all wrong. I said, No, you're all right. Here's the next version. And because everybody loves being listened to and influencing the output and conclusions of things, I sort of became a bit of a mascot for the industry," he muses.

His vision and foresight powered his adaptability. In many ways, risk-takers in the renewables industry acknowledge from day one of their venture that change is inevitable and will disrupt their original business model.

Michael admits he, "Suffered a very substantial sort of loss of confidence" before starting the new venture. Total revenue for NEF in 2004 ended up being two thousand pounds, and he saw the two hundred thousand pounds in his personal bank account quickly diminishing.

Sitting back a few years later at REFF-Wall Street in April of 2007, he listened to CEO after CEO speaking to the coming short supply of silicon, which had spiked to $450 a kilo as well as the problems around solar panels being priced based on the principle of the German feed-in tariff, which was declining at a trivial 5 percent a year.

In addition, oil was peaking at twenty-eight dollars a barrel. Going back to the knowledge he had acquired at his first job out of Cambridge with a small consulting company, he knew an experience curve was missing in the discussion.

Michael had a sense of what was going on with the energy sector. "I knew that for every doubling of cumulative volume, you get the same price drop and in every industry, that goes on forever. Commodities can go up and down, but the value and price just go down."

Drawing a chart on a notepad in front of him, he pulled together one of the first experience curves for solar. As he continued to absorb information from the conference, later combined with initial data from a consultant named Paul Maycock, he concluded that the price of solar would go flat between 2004 and 2008 and then would crash in 2009.

With concept in hand as well as a curve showing the necessity to inform an industry of an impending crash, Michael went back to London to further define his business model, identifying revenue sources, a customer base, and a product. He knew that to build a successful start-up, he needed to take a full 360-degree assessment of the opportunities in the clean energy sector as well as hire the best talent he could afford along with "a lot of luck." To Michael, the time to launch the concept was then since he sensed that the ice was cracking.

Employing an armada of interns who were interested in the sector and paying them ten pounds a day, he launched this ambitious effort. Having been burned by the venture capital industry during the dot-com bust, he committed to putting in seven thousand pounds a month of his own money, as well as leveraging his brother-in-law and his American Express Credit Card to provide initial seed capital.

In reflecting on his motivations in starting up NEF, Michael knew he was not a big believer in transhumanism but was driven by the thought that we, as a society, should try and leave the planet a better place than we found it.

According to the *Dictionary Britannica*, transhumanism is a philosophical and intellectual movement, which advocates for the enhancement of the human condition by developing and making widely available emerging technologies that can greatly enhance longevity and cognition of the human race.

And indeed, unbeknownst to Michael as he was starting out in renewables, he was a transhumanist. He confesses that even to this day, he has a very "strong belief that we should try to be good to each other and good to the planet."

Growing up in the 1970s and 1980s, he in many ways felt that those having grown up in that era were taught to believe "stupid things" and that part of his journey was to deprogram a lot of those concepts or biases that were either unacceptable or unattainable. A key here for him was to shed those attitudes that in many ways were barriers in the way of him achieving his goals and adapting to the reality of the situation.

Michael's advice in starting up any venture is: "Don't take 'no' for an answer. You have to work pretty darn hard, and there are some sacrifices and costs, but you can do a lot more than you think."

He shed his "can't do" attitude, which helped him immensely in those early days. The interns he had assembled worked on everything from logo creation to strategic product development. He created a community around him that was adapting to the new sector just as much as he was. He had inadvertently created an oak tree for like-minded young professionals interested in "cracking the nut" of renewables finance.

Using sports and paying attention to his mental health, Michael kept climbing, skiing, and volunteering at the St. Mark's Hospital Foundation, keeping his physical and mental stamina up. To him, the entrepreneurial foundation of any new venture is based on the CEO's personal ability to keep going while being the first person in the office every morning and the last person in the office. According to Michael, "If you stop, your team definitely stops."

The NEF team grew closer and developed an "asshole-free zone," with Michael firing anyone who in any way brought down morale around their mission. "The ability to do that while growing revenues over 500 percent a year and working on the right issues in the world—I don't know that I'll ever professionally be able to recreate that," Michael recollects.

Together, they started focusing on data speaking to every investment they could find in the clean energy sector and "extracted their DNA"—as Michael refers to it—tracking the investor on the deal along with the debt provider, creating structured data sets.

Though customers were starting to sign up, Michael quickly began to understand that for these customers, the subscription wasn't the largest investment. The largest investment was the time spent analyzing the data the New Energy Finance team had pulled together. So, supplying data, was not going to be the ultimate product.

Enter in Ken Bruder, a Harvard Business School alum, who told him he needed an insight service to provide information on the clean energy sector's market size, price forecasting, supply chain analysis, and much more.

Michael had throughout his years of business understood the value and necessity of listening to mentors like Ken, especially in acknowledging the need for humility when accepting feedback and in some instances, pausing and retooling a concept or product.

"You've got to really not kid yourself. Sometimes the product just doesn't do what it's supposed to do, and you have to be okay with re-strategizing," reflected Michael.

He latched on to Ken's advice, having identified the "right product" that had been evading him and adapted to a new course in strategy. He started a series of newsletters, targeting those initial customers the team had gathered. By 2005, he had hit a mega trend with the NEF platform having become the "go-to" data resource needed by countless new entrants into the market ranging from the investment sector to technology providers and energy giants. These wins assisted Michael in courting angel investors he trusted in further growing out the platform who believed in the impact they were making.

And the NEF team's work was having impact. Investors and technology suppliers armed with NEF forecasts behaved and invested differently as a result. Having time to diversify their product, NEF in 2006 did another "industry first" as it started to look at the effects of a levelized cost of solar on the market. Indeed, Michael and the team were taking a reputational risk in issuing many of these types of forecasts, but the popularity of the NEF Briefings took off.

"Looking back and saying there's going to be a crash and not an orderly price drop—even potentially an overshoot— was brave. You know, it just feels less brave now in retrospect," reflects Michael.

He saw the industry taking off despite supply and demand challenges. Working with Ken Bruder, from 2007 his team developed industry-shaping presentations, including a word slide that showed the net zero transition taking decades, costing trillions of dollars, and requiring extensive public policy paired with private money.

His thoughts were that the modern renewables industry would comprise less than 1 percent of the global electricity supply in 2008 and, while it would ultimately reach

15 percent, that would take his entire career. Instead, we're already seeing these figures today. Michael's response: "We are creating that reality."

Michael proved his foresight time and time again in creating NEF. He literally checked every box of what Leadership Strategist Jeff Boss in a *Forbes* piece ascribes to an adaptable leader, specifically that these types of risk-takers: (1) experiment, (2) see an opportunity where others see a failure, (3) are resourceful, (4) think ahead, (5) are curious, (6) stay current, on knowledge acquisition (7) see systems (i.e., the forest from the trees), and (8) open their minds to new possibilities.

Strategist J. Peter Scoblic would agree with Boss, and in his piece in the *Harvard Business Review* adds that strategic foresight implemented by successful leaders does not necessarily aim to predict the future but to help organizations envision multiple futures in ways that enable them to sense and adapt to change through scenario planning.

For organizations to continue to thrive, their leaders continuously challenge their teams to imagine a variety of futures, identify strategies that are needed across them, and begin implementing those strategies now.

Michael's foresight played a fundamental role is his adaptability to the evolving renewable energy market and finding the right time and right market conditions to launch NEF.

"The timing was incredibly lucky because if I had started New Energy Finance in 2003, instead of 2004, I would have run out of oxygen. And if I had started in 2005, instead of 2004, there would have been twenty-two Michaels all running around, all straight out of McKinsey, with a better network than me and you know, better mentors than mine, which in 2003 were essentially nonexistent," states Michael.

His foresight continues to stand out immensely. In marketing, it's all about forecasting the customer's needs and positioning the product you're working on to fulfill the customer's needs. Michael was always at the forefront of understanding the customer—in this case the broader renewables industry—and anticipating the needs of an organically growing industry. He understood the industry's growth would not be linear, and in that realization he found a business model that would thrive.

Before Ken Bruder's arrival in 2007, the business had two paid products: a robust investment database "desktop" service along with its corresponding digital NEF Briefings. That April, NEF launched its hugely successful NEF Insight Service, which slowly took over as the primary paid product for NEF. This product got the attention of the Bloomberg empire.

Seeing the value NEF had created in the clean energy research and industry knowledge acquisition markets, Bloomberg approached Michael to negotiate acquisition of the company. In April 2009, he decided to sell the company for a rumored eighty-five million dollars and the final deal closed on December 9, 2009. The angel investors Michael approached back in 2006 made eight times their original investments in just over three years—something unheard of in the angel investment sector even to this day.

Remembering back to when Michael was scribbling thoughts down on his notepad at REFF-Wall Street in 2007, he mentioned a potential solar pricing crash in 2009. Guess what? It indeed crashed by 40 percent and then crashed again by 40 percent in 2010, according to BNEF.

Michael's foresight had yet again proved correct and continues to prove correct even to today.

CHAPTER 12

Servant Leadership

Andy Karsner

"Politicians and industry have the climate change problem identified in spades, but not enough problem-solving going on. The government's decoupling of the price of oil from the price of renewable energy in purchasing energy is a significant watermark, and I encourage more people inside and outside of the political process to become agents of disruption," declared Andy Karsner during the Phase II of Renewable Energy in America Policy Conference on Capitol Hill.

In 2005, during the initial months of President George W. Bush's first administration, Spencer Abraham, the first Arab American to have held the post of Secretary of Energy had served the president well and was succeed by Samuel Bodman.

Secretary Bodman held a PhD in chemical engineering from MIT and had also served as a professor and departmental director at the university. He also knew the DC political landscape well, having previously been the Deputy Secretary of the Treasury and Deputy Secretary of Commerce under the president.

Despite Secretary Bodman's connection to the traditional energy sector, the president was becoming more and more intrigued by conversations and news coming from the corporate energy sector about the potential for renewables.

Following 9/11, President Bush understood America's addiction to oil and the disastrous ramifications it had on the future of the US all too well. Bush was a proud Texan, remaining in touch with colleagues and friends in Texas, who were starting to get into the wind business and were making some serious money.

To those of you that remember, this was the era of T. Boone Pickens and the Pickens Plan. Best known as an oil man and corporate raider for his runs at Gulf Oil, Unocal, Pioneer, and others in the 1980s, T. Boone Pickens focused most of his time in the 2000s on managing Dallas-based hedge funds and pushing his Pickens Plan to boost adoption of wind, solar, and natural gas.

The plan, which reached its peak between 2007 and 2008, was a multi-million-dollar advocacy campaign run by T. Boone Pickens to reduce the US dependence on foreign oil. The plan called for replacing oil imports with domestic energy by harnessing wind power for electricity and using natural gas as fuel for vehicles.

The president was looking at how to organize and transform DOE to make it more commercialization focused. It was time to bring in some corporate private sector experience into the US Department of Energy, and President Bush had an idea.

The concept of being a "public servant" is today often associated with a bygone era when working in government was seen as a privilege, an honor, and an opportunity to serve the American people. Being a public servant was seen as answering a call to serve others, citing President John F. Kennedy's statement during his January 20, 1961, inaugural speech, "Ask not what your country can do for you. Ask what you can do for your country."

From my education at Georgetown University, another motto that still echoes is the Latin phrase "Servitium pro aliis" or "Service for Others." I had the fortune of interning briefly in Ari Fleischer's Office in the White House under the second term of the George W. Bush Administration, and every day was filled with awe walking in past security and the media and straight into the West Wing's press briefing room.

This awe was matched with a sense of gratitude for the opportunity, which randomly was presented as a temporary "fill-in" internship to me when I worked at the School of Foreign Service (SFS) Dean's Office at Georgetown.

Looking around at the staff in the West Wing, I sensed dedication, and a commitment to serve the American people, with no one taking for granted the limited time they had to create an impact in four—or in this case—eight years.

Many have asked how we reignite the passion and longing to serve among not only public servants but also those in all forms of business. Many seek the common denominator that made the leaders of the past so much more inspirational than the leaders of today. The art of servant-leadership needs to be promoted again.

In his book, *The Servant as Leader,* Robert K. Greenleaf is credited with being the first in utilizing the phrase "servant-leader," stating, "The servant-leader is servant first... that person is sharply different from one who is leader first, perhaps because of the need to assuage an unusual power drive or to acquire material possessions."

Greenleaf goes on to state that the servant-leader puts others first, which means they are not the sole leader with power but rather a power-sharer. By putting the needs of others above their own, they are able to empower their team to grow, develop, and strive to achieve success for the entire team.

Another organizational leadership expert, Peter G. Northouse in his book *Leadership: Theory and Practice,* goes on to ascribe ten characteristics of servant-leadership: (1) listening, (2) empathy, (3) healing, (4) awareness, (5) persuasion, (6) conceptualization, (7) foresight, (8) stewardship, (9) commitment to the growth of people, and (10) building community.

In describing these characteristics, it becomes evident that servant-leadership does not only apply to an internal business organizational structure, but it also has ramifications and impact in how we interact in greater society.

"In business, servant-leadership would start with your customers and ultimately involve serving society through the good work you're doing on behalf of your customers," states Bill George, a professor of management at Harvard Business School and the former CEO of Medtronic, a large medical technology company.

With this call for us all to be servant-leaders, we need to be open to answering and implementing in the advancing of the next phase of renewable energy leadership. As a renewables industry, we're lucky that many in the beginnings of their careers heard the call to advance renewables to serve others. One of those who answered the call was Andy Karsner.

Andy was an entrepreneurial venture capitalist. Being born into a multicultural Swedish Moroccan family and to a father who served in the US military, he had big ideas and enjoyed being a globe trekker from the very beginning. He was also known for having the knack of being in the right place at the right time.

After a few years of the family having relocated from California to Texas, during which time his father was deployed abroad for clandestine operations for the Cold War, Karsner became the first person in his family to graduate from a four-year institution.

He attended Rice University, where he majored in political science and religious studies, and he graduated with honors in 1989. He then went on to study international Finance at the University of Hong Kong, earning a master's degree in Comparative Asian Studies.

His newly acquired Texan roots, combined with his sense of adventure led Karsner to embark on his journey in energy in 1988, as an energy analyst and project manager for international development at Tondu Energy in Houston, Texas. Tondu was a pioneering independent power producer. He was then hired in 1994 as a Senior Development Manager for Wärtsilä Energy, based in Hong Kong.

At Wärtsilä, he was attracted by the energy innovation going on at Enron Renewable Energy Corp—led by the visionary chairman and CEO, Bob Kelly—which his mentors had put him in touch with. Andy was skeptical of the ability of "windmills" to scale but was intrigued by the renewables technologies being innovated as well as the leadership he saw coming to the helm of the budding industry.

Bob Kelly had made a name for himself almost a decade before, having overseen the purchase of Zond Systems, America's largest remaining wind company in 1997, along with German manufacturer Tacke Windtechnik in 1997.

Zond was created in 1980 by Jim Dehlsen, who went on later to form Clipper Windpower in 2001.

Upon completion of the acquisitions, which were the first-ever of their kinds in the wind industry, Kelly stated, "Our purchase of Zond is part of a broader strategy, a vision. We believe renewable energy resources will be capturing a larger and larger share of the power market within the next twenty to twenty-five years."

Zond had developed the largest wind turbine model—the 1.5 MW Tecate—which years later in 2002 was bought by GE Energy as part of its acquisition of Enron's wind business during bankruptcy proceedings.

The remnants of Enron's wind business were the only surviving US manufacturer of large wind turbines at the time. GE Energy increased its engineering and support for its wind division afterward and doubled its annual sales to $1.2 billion in 2003. It later went on to be an active participant in the "West Texas Boom" of the later 2010s based on the momentum they got from the acquisition.

Bob was keen to recruit talent and build a team that shared in his ambitions and had turned around and recruited H. David Ramm from United Technologies Corporation to assist him in growing out Enron's renewables portfolio.

A previous Army officer, David excelled at Enron as managing director of Enron Renewable Energy Corporation, later climbing the ranks to become president of Enron Wind Corporation. Many of you may recognize his name today, given his roles as Managing Partner of Dymar Development, chairman and CEO of BrightSource Energy, Inc., and Chair of the University of Houston's Energy Advisory Board.

Under Bob and David, Enron Renewable Energy Corp became the second largest provider of photovoltaic technology in the world, having formed partnerships with Japanese companies to install PV on rooftops in Japan. The company was also very active in the PV markets in the US and India. In addition, Enron was pushing hydro and geothermal technologies, and their ambitions were growing.

Enron made a key investment in the solar industry, investing in the largest solar manufacturer at the time, named Solarex. Solarex was a fifty-fifty joint venture between Amoco and Enron Oil and Gas, and at its peak had annual revenue of fifty-eight million dollars in 1998.

Cofounded in 1973 by innovators and physicists Peter Varadi and Joe Lindmayer, Solarex was the first to develop the utilization of solar cells for terrestrial applications in the US. They decided solar power could be brought "from space to Earth" by bringing down the costs, and they founded the company to industrialize the manufacturing of polycrystalline silicon solar cells using their own patented process.

The plan for Enron was to engineer Solarex into a viable solar enterprise and flip it to garner maximized value. The plan succeeded and in 1999, Solarex sold its remaining shares to BP Amoco PLC, who paid forty-five million dollars. Solarex then became BP Solar.

Bob and David's leadership had inspired Andy. He also was learning a lot about the energy development business at his current job at Wärtsilä, under the leadership of Ian Copeland, who was the Managing Director of Wärtsilä Power Development (Asia). Ian was well-known as a developer for his accomplishments that included the privatization of, and two-billion-pound financing for, the infrastructure portion of the London Underground system.

He also was lauded with the delivery of the world's largest solar thermal power project at that time. In addition, Ian developed and financed greenfield power plants in Indonesia, Pakistan, China and the Philippines and buildings and energy-related assets used to support the US federal government. Andy had learned the ropes of the energy development business globally and was ready for his next adventure.

In 1999, he moved to London, where he became founder and CEO of EnerCorp, a wind-power development firm and international agent for Danish wind leader Vestas. Andy was involved in project development, management, and financing of energy infrastructure globally and had worked on projects spanning developing the solar market in Vietnam to advising the country of Morocco on becoming a net exporter versus net importer of energy.

His drive helped him achieve a milestone in 2002, having led EnerCorp to win a global competition to develop the world's largest private wind farm outside the United States. His successes at EnerCorp led him then to establish Manifest Energy, a firm committed to energy development, investment, and financing.

At this time, his work with utilities, manufacturers, and other clients put him in the crosswinds of political dialogue. Having become an ardent supporter of renewables and given his global work, he had seen the transformative socioeconomic impact renewables could make in the US.

In late summer of 2005 Andy received an unexpected call. He was being invited to engage in conversation with the White House on a call-to-action President Bush wanted to make to curb the country's addiction to oil. That conversation then led to a request to accept being nominated to become the next Assistant Secretary for Energy Efficiency and Renewable Energy at the DOE.

"I was actually reluctant to do it," Andy recalls. "In fact, I got a little bit spooked about how long and deep the FBI clearance process was. I was like, what am I doing? I'm jumping off a cliff and giving up my privacy as a person."

A few days after the call, Andy found himself at the Reagan Building as part of the initial interview process. Given the size of the renewables industry, the word was already out on DC streets of Andy's nomination process. Sitting down with two industry colleagues he respected in the basement of that building, he wanted to get their insight and advice on taking on the role.

In front of him was Scott Sklar and next to him was John Mizroch, who would become Andy's right hand at the DOE, as the Principal Deputy Assistant Secretary for Energy Efficiency and Renewable Energy.

John was a career federal government executive who knew the processes, systems, and ways to get things done in bureaucratic Washington, according to Andy. He began his career as a Deputy Assistant Secretary of Commerce for the Trade Development and International Trade Administration at the US Department of Commerce in 1987.

Two years later he took a position of an advisor to the minority at Joint Economic Committee of the Congress, where he worked until 1990. In 1993, he was appointed president of the Environmental Export Council. John then worked as president and Chief Executive Officer at the World Environment Center from 2000 to 2006. For a brief time, he served as acting Assistant Secretary at EERE while Andy was going through his nominations process.

Weighing the economic, political, and opportunity-cost risks to his growing business, Andy asked for their advice—as all servant-leaders do—needing an affirmation that what he was going to embark on would help society as a whole as well as continue his family's dedication to serving others.

"People wait their whole life for this call, Andy," chimed in John. "People dream about serving sometimes for the wrong reasons, sometimes for the right reasons. It is so rare to have the idea that the president of the United States will listen to your perspective, yet this man is searching for a new direction. And they're asking you to lead it."

John's comment stirred something in Andy. Andy was the first male member of his family since the time of immigration who had not worn the uniform of our country. "And we were in the middle of the war with Afghanistan post 9/11 and I was being asked to serve," recollects Andy.

His mind was made up and he was going to do it. Andy's confirmation went smoothly, and he was unanimously confirmed by the US Senate on March 16, 2006 and sworn in on March 23, 2006. During this swearing-in testimony, Andy stated, "Our liberty comes with responsibilities and our opportunities imply an obligation" and that serving in this position would be a demonstration of "an appreciation of service above self."

Andy relied on the lessons from Bob, Dave, and Ian as well as the oak tree of support—both within and outside DOE. He quickly assembled a team that would take on the president's mission for energy independence. Key to that mission was to expedite the commercialization of renewables to curb the US dependence on foreign sources of oil.

Andy was given the task to oversee approximately two billion dollars in annual federal funding that would be focused on applied science, research, and development portfolios for clean energy technologies.

On the advice and introductions of Scott and John, Andy met with an informal kitchen cabinet of sorts, garnering advice on how the government could facilitate greater collaboration between the public and private sector.

He most importantly wanted to do so without the government getting in the way or being seen as choosing between "winners" and "losers" in the industry. As a servant-leader, Andy knew transformative leadership had to take a humbler approach—one that listened before leaping into action.

These meetings put him in touch with a variety of leaders—ranging from Amory Lovins of the Rocky Mountain Institute and Dan Reicher who was his predecessor in the Clinton Administration to Michael Eckhart of ACORE.

He also made the trip up to New York to headline at a finance conference named REFF-Wall Street, meeting with John Cavalier, lead for Credit Suisse's renewables practice, and Alan Waxman, a Partner at Goldman Sachs and Chief Investment Officer of its largest proprietary investing business, to better understand what incentives were needed to finally make renewables attractive to investors.

"I consulted everybody," recalls Andy.

He approached each conversation similarly. "We're at war. How do we win?" Reminiscing on the innovative advances made during the Manhattan Project, he saw himself advancing an agenda that was similar to developing a secret weapon.

However, to Andy, it was no longer, "Guys and scientists on a chalkboard working on this in New Mexico. Instead, we have entrepreneurs in garages, and we've got to support them in getting out of oil."

Andy lunged into the job, achieving political wins for the administration in helping assemble significant bipartisan coalitions to implement and enact a number of policies, including the earth-moving Energy Policy Act and The Energy Policy Act of 2005 that forever changed US energy policy by providing tax incentives and loan guarantees for renewable energy production.

In addition, the act also stipulated an increase in ethanol production to be blended with gasoline. His focus continually was aimed at killing our addiction to foreign energy while garnering US leadership in renewables. Andy's charismatic way of engaging leaders in bipartisan dialogue resulted in the successful passage of the Energy Independence and Security Act and the America COMPETES Act. All these policies remain in place today.

During his conversations with lawmakers, "I never talked about energy independence," says Andy. "I would start speeches by scoffing at the term and saying we're always going to be an interdependent world. To me, the independence that matters is being rid of the leverage that our enemies can put upon us. It's not about energy independence. It's about our addiction."

In essence, Andy was realizing we needed to resolve the addiction, and like any addiction, we needed to admit we have one and seek out a path to recovery and rebirth as an energy industry.

A slow collective realization had started to take hold of many of those leaders—the realization that having an addiction provided a needed challenge for the energy industry, as well as leaders who could guide us through the beginning of our recovery.

Cue in the need for greater servant-leadership. Andy utilized his platform and in true entrepreneurial style proceeded to flip the traditional power leadership model and put external stakeholders and his team members front and center as those who knew how to get us on the road to recovery.

By taking his position and making himself accessible to key renewables stakeholders, as well as his internal team, he empowered—and indeed challenged—us all to do better. Though servant-leaders are dedicated to the growth and development of their coworkers, they do not view themselves as a pure servant.

That's where the leadership part of the phrase comes in. Servant-leaders cannot elicit pro-activeness in their team members or substitute a lack of drive among them. That is up to the team member alone.

Andy viewed his servant-leadership role as one that was in charge of clearing the hurdles of politics and ironing out outdated processes that would allow his team to advance the public-private partnership between the US government and private sector. His actions also made the EERE's commercialization program open to allow cross-pollination of efforts and create a place for all of us to actively participate in government's role.

Servant-leaders, according to Art Barter, Founder and CEO of the Servant Leadership Institute, are "Serving instead of commanding, showing humility instead of brandishing authority, and always looking to enhance the development of their staff members in ways that unlock potential, creativity and sense of purpose.

The end result? "Performance goes through the roof," says Barter. Indeed, the success around DOE's Commercialization Program was just beginning.

On August 29, 2008, the commercialization team got their first win with the announcement of DOE's new seven-million-dollar fund called the Technology Commercialization

Acceleration Program. The program was focused on the "technologies on the shelf" at our national laboratories across the country and figuring out how to get them to market and scale.

John Mizroch on the occasion of the announcement stated, "This funding for technology commercialization will help bridge the gap between our labs and American consumers." He added, "It is absolutely critical that we move clean energy technologies to market at a rate and scale that is commensurate to the magnitude of the problem. Our environmental well-being, national security, and economic competitiveness are waiting for it."

Andy had his sights on having the government and Silicon Valley collaborate on the scalability of renewables technologies. He needed to add a team member who was young, finance-savvy, and a doer. That came in the form of Michael Bruce, who became the Senior Advisor for Finance on the DOE commercialization team. Michael was a former Stanford swimmer who had built a strong reputation for his financial acumen and networking skills at Credit Suisse.

In interviewing Michael, Andy used an old McKinsey interview question: "I put you in a room. You've got a telephone, nothing else. What do you do?"

Michael, without taking a pause said, "Well, that's easy. I call over to Silicon Valley and I set up a trip for you to meet with every venture capital firm to change white papers into business plans. We then lead an effort to make sure that every early-stage investor and high risk-taker in America knows we're on their side so we can accelerate capital into clean energy."

With that, Andy hired Michael. By the next week, Andy and Michael were on a flight to San Jose.

The team engaged in commercialization conversations with key Silicon Valley leaders such as venture capital giants Vinod Khosla, cofounder of Sun Microsystems, and the founder of Khosla Ventures, and John Doerr of Kleiner Perkins. Andy and the team had a simple message.

For the first time ever, DOE had significant funding and agreement from the national laboratories to pursue renewables as well as a mandate from the president to unlink ourselves from oil and gas.

"We want business plans and to embed your entrepreneurs in our national labs," pitched Andy. He went on to say, "We will have those national labs validate and verify the stage gates of the technologies you're investing in and DOE is going to sprinkle nondilutive equity money into your deals to super-size them."

John Doerr loved the idea. At that meeting, he and Andy agreed to coining the new term "cleantech" for the industry as well as starting an "Entrepreneur in Residence" program with DOE that Michael had recommended. They had accomplished charging up the national laboratories with a new mission and had infused Silicon Valley's connection to the innovation going on inside.

And what was the result once they got back to DC? The creation of DOE's very first Technology Commercialization Fund, which still exists today.

CHAPTER 13

Dedication

| Steve Chalk | Carol Battershell | Drew Bond | Wendolyn Holland | Amy Chiang |

"We all took general risks starting out in the industry, as a certain type of dedication was required in order to stare directly at the challenges, take a deep breath, and then commit to going all out," reflects Wendolyn Holland upon the start of her journey in renewables.

She went on to add, "The telling of the story of what really happened in the early days of renewables is important, so that we—and future generations—understand our path on the arc of history and why it's vital to continue to take risks."

Andy Karsner's success with the DOE's Commercialization Program would not have been possible without a strong dedicated team behind him: Steve Chalk, Carol Battershell, Drew Bond, Jennifer Owen, Wendolyn Holland, Amy Chiang, Brad Barton and Jacques Beaudry-Losique. I've had the fortune of remaining in touch with Carol, Drew, Wendolyn and Amy, and we had a chance in the spring of 2022 to catch up on old memories.

It's important to acknowledge that in every conversation I had, Andy's style and ability to inspire came up as one of the first recollections to the success of the team. His approach garnered traction with internal DOE employees, as well as energy-sector stakeholders ranging from investors on Wall Street to start-ups in Silicon Valley—all individuals who were observing a steady and clear stream of support for renewables issuing forth from Washington.

The Bush Administration was known for its team-oriented culture, and the DOE commercialization team personified that daily. According to organizational author and blogger Georgina Guthrie, team-oriented staff members can focus on motivation rather than daily completion of tasks and deadlines.

By avoiding complicated hierarchies and tedious processes, and by learning to trust each other, teams are nourished and empowered to achieve an overarching goal. Successful teams share a strong sense of communal dedication to an ideal or organizational mission.

Andy and his team checked all these boxes without trying to as they lived, ate, drank, and slept the concept of quickly scaling up and commercializing renewables.

I was approaching my midway career point at ACORE as Andy was pulling together this first ever commercialization team at DOE, and I remember how he was able to rally support from ACORE's CEO Mike Eckhart. Over and over, he stressed the importance of shared dedication to bring about industry transformation through truly innovative public-private sector collaborations. These partnerships focused on commercializing and scaling technologies that were stuck in entrepreneurial garages or national laboratory shelves for way too long.

Individuals and teams dedicated to the US renewable energy industry at that time were challenging the traditional energy status quo in their efforts to create a more sustainable world. Whether they were aware of it or not, Andy and his team were about to rewire society's foundational understanding of sustainability, and in doing so provide a possible solution to the seemingly intractable challenges of climate change.

A key point Andy kept making was that we needed dedication to persist and push the envelope on scaling public-private partnerships. Each new member Andy brought on shared one common trait—a dedication to the overall mission.

The definition of dedication within the *Oxford English Dictionary* is the quality of being dedicated or committed to a task or purpose. Coming into the renewables industry requires dedication to achieve aspirations that are much greater than ourselves.

This type of dedication was especially exhibited by early industry risk-takers, who came armed with a commitment to transforming organizational and societal structures for the better in tackling climate change.

According to David Lancefield and Christian Rangen in their essay in *Harvard Business Review*, "A substantive and irreversible shift in an organization's identity, value system, and capabilities requires three difficult acts: Developing a deeper sense of purpose that guides strategic decisions and shapes the workplace culture, repositioning the core business, and creating new sources of growth."

Andy understood that his ideal team would include both DOE insiders and outsiders in order to skillfully navigate the energy political landscape, effectively advance technologies to scale, and profitably hit public markets.

His dream team would attract thought leaders and others from outside Washington who could understand the dizzying potential of the outward-market-facing and entrepreneurial nature of the proposed commercialization program.

Some of the first members of the team included Steve Chalk and Carol Battershell. Steve had been quietly working in cities across the US on innovative ways to mainstream renewables, employing technologies ranging from fuel cells to solar energy.

He was also instrumental in establishing the US ethanol industry based on nonfood biomass sources, converting wood and wood waste from pine forests and mills into millions of gallons of ethanol per year.

Carol Battershell was a twenty-five-year industry veteran who held a wide range of senior executive roles at BP. Over the years she has focused on alternative fuels and renewables including remediation, compliance, strategy, policy, research, operations, procurement, and commercialization.

Carol's interest in renewables began at a tender age. Having grown up near Cleveland, Ohio, pivotal memories from that time include: "A river that burned—that made a great impression on me—and a Great Lake so polluted that some summers we were not allowed to swim in," she shared with me.

While a vice president in BP Alternative Energy, she took an MIT class with Peter Senge on how all sectors—industry, government, universities, media, NGOs—will need to work together to solve the world's largest problems, including those such as climate change and poverty and water scarcity.

Throughout her career, she felt the call to serve greater society. Near the end of her tenure at BP, she helped convince the company's board to invest eight billion dollars in a new division called BP Alternative Energy.

Leaning in for a big career change, she applied for a job with the US federal government, working at the interface between industry and government to bring clean energy technologies out of the national labs and into the market.

"Coincidentally, I applied at a time when Assistant Secretary Karsner was working to bring more private-sector executives into the government," she recollected.

At DOE, she had made a name for herself, where she led multi-billion-dollar technical programs, ran the Energy Efficiency and Renewable Energy Field Operations Office, and was a key contributor on two multi-agency energy-policy reviews.

The team expanded around the spring of 2007 and with it came the additions of a few specific movers and shakers dedicated to expanding renewable energy technologies in the marketplace, including Senior Advisors Drew Bond and Wendolyn Holland, Senior Advisors for Commercialization; and Amy Chiang, Director of International Affairs for EERE. Though all three were located in different offices across DOE, they banded together, dedicated to building out DOE's commercialization team.

The coalition of the willing had come together to support DOE's objectives.

Drew Bond and I met virtually late in the fall of 2021 to look back at his career focused on public service, his extensive background as an entrepreneur—he had just wrapped up a home-builder start-up—and his DC experience as Chief of Staff at the Heritage Foundation under Ed Feulner.

He also had energy bona fides, having served as Chief of Staff to former Oklahoma Corporation Commissioner Denise Bode, and as a Legislative Assistant to former US Senator Don Nickles from Oklahoma. Thinking back on his role with the commercialization team he reflected, "I knew nothing about renewable energy, but I knew that as a country we really needed to diversify our energy sources and get off of the international and geopolitical merry-go-round we'd found ourselves on."

Wendolyn Holland and I had remained in touch since our time in Washington, and she was known in the industry as a connector, having excelled as a consultant assisting firms in the sustainability and clean energy industry to achieve their business goals and interface with federal and state agencies.

Catching up with her virtually, we talked about her beginnings in renewables as a consultant, when she had also backed many initial start-ups in the sustainability sector, some of which took off while others failed. "One had to be willing to embrace failure, take a hit to one's reputation, and then dust your boots off and move on the next one," she told me. "Vision and hope keeps us coming back to support start-ups."

Amy Chiang brought years of experience in renewable finance and infrastructure deployment to her position but also had political acumen on her résumé. Coming from a Chinese American family, she had previously cofounded an advisory firm focused on energy and infrastructure investments in China.

She'd gained policy experience during her time as Deputy Chief of Staff for US Congressman Sherwood Boehlert, chairman of the House of Representatives Committee on Science. Her background also includes working on Wall Street as a foreign exchange derivatives trader and as a management consultant with Booz Allen Hamilton.

"It was an exciting time to be working at the DOE and partnering with international stakeholders," recollected Amy. "We believed public policy in tandem with political leadership could help drive wider acceptance of renewable energy and energy efficiency technologies globally. The US has a strong innovation story and track record on deployment of these technologies that we wanted to share. We also wanted to work closely with other governments to exchange lessons learned and broker international cooperation to help foster the global energy transition."

It was still 2007 and the commercialization team continued to grow with the additions of Brad Barton, Director of Commercialization and Deployment, and Jacques Beaudry-Losique, Deputy Assistant Secretary Renewable Energy. As individuals, and as a group, they were dedicated to responding to the call to commercialize renewable technology *fast*, and to accelerate innovation, especially given the geopolitical backdrop of the United States' involvement in Iraq.

"We weren't there to get a government paycheck or bolster our résumés," Drew said. "We were all about speed, scope, and scale."

After each whiteboarding brainstorming session, Andy challenged the team to think bigger. And they did, instigating DOE's Technology Showcase—a crosspollinating effort across DOE's departments and national laboratories to both showcase the best innovation the country had to offer, and match it with private-sector investment.

The initial pilot program in 2007 brought twelve investors to the table. A year later, the showcase attracted one hundred investors along with a diversity of other non-venture-capital investors, including universities and start-up incubators. The program had successfully bridged the divide between public-private financing of renewables.

Much like how individuals break through barriers of fear and institutional barriers as described by author Gay Hendricks in his book, *The Big Leap* by "removing the upper limit problem that prevents you from living up to your full potential," the commercialization team was dedicated to their mission and empowered to take risks daily.

Though their daily risks were calculated ones, they were tied to the belief behind the impact technology commercialization would have on the global market.

Another impact-filled initiative was a collaborative effort involving dozens of dedicated national labs team members to create a database with the national labs to help companies find clean energy technologies to license and develop.

Carol, Wendolyn, and their team set up an internet portal to allow searches of the thousands of technologies all the DOE national labs had developed and patented. Their Tech Transfer colleagues in the Labs designed and wrote hundreds of one-page marketing brochures for the inventions thought to be most likely to be of interest to a wide-range of industry.

"Prior to this publicly available database, the onus was on a company to guess which individual labs to contact in order to ask multiple researchers what technologies might be available for commercialization," noted Carol.

What appeared to be a simple database exercise had immense ramifications, opening the closets throughout the national labs and allowing sunlight to shine on innovations that had been tried, tested, and underutilized by an industry hungry for more technologies.

In summarizing his view on the team and the magnetic dynamism that surrounded them, Andy said, "We had a common vision to overcome the status quo of traditional energy, which was not secure and was deleterious to our future and corrosive to our economy, which was bleeding our environment. All of us knew better ideas were out there, and we had the capacity to tap into those ideas."

In a *Harvard Business Review* piece by Robert S. Huckman and Bradley Staats titled, "The Hidden Benefits of Keeping Teams Intact," a central thesis is that performance improves with a team's familiarity with one another and their associated strengths and specialties. According to the authors, the greater the amount of experiences and trust individuals share in approaching challenges can only help to increase the influence on a team's performance and success.

The DOE commercialization team personified this thesis, not only through their dedication to one another but also to the commercialization mission and to the overarching industry. This dedication would be a driver for the rest of their careers. The team was dedicated to solving the problem of connecting viable technologies to available sources of financing and de facto acted like a lean start-up in their own right. As with any true start-up, dedication was key.

The collective dedication brought to the program by Steve, Carol, Drew, Jennifer, Wendolyn, Amy, Brad, and Jacques, and others resulted in another byproduct—a high-performing team.

From setting clear goals and organizational priorities to clear communications with renewables stakeholders to trust each other and push each other to be accountable and daring, they personified the characteristics of what organizational consultant Kristen Ryba defines as the eight characteristics of high-performing teams.

These characteristics included having clear goals closely tied to being dedicated to organizational priorities and understanding how their work fit into the overall mission. In addition, they defined their roles and responsibility, communicating clearly while remaining focused on wins for the team and for the DOE.

Most importantly, the team practiced continuous learning, scoping, and studying all the technologies that were sitting on the shelf at DOE and at the national laboratories.

The DOE commercialization team personified the tenets of this thesis, not only through their dedication to one another but also to the commercialization mission, the industry, and the betterment of the world. Parallel and mutual dedication had driven them up to this point and would inform them for the rest of their careers.

Connecting viable technologies to available sources of financing turned out to resemble a lean start-up, and with each win the team's commitment to each other and the product intensified, and they worked harder.

The team had garnered two more significant wins during their remaining years at DOE. The first was the launch, with Governor Linda Lingle, of the Hawaii Clean Energy Initiative. Focused on reducing Hawaii's oil dependence by 70 percent through substitution of renewable energy and energy-efficiency measures, in many ways this initiative was the first state-wide pilot microgrid demonstration undertaken in the US.

The second was championing through the Technology Commercialization Fund (TCF). The TCF is a nearly thirty-million-dollar funding opportunity that leverages the R&D funding in the applied energy programs to mature, promising energy technologies with the potential for high impact.

It uses 0.9 percent of the funding for the Department's applied energy research, development, demonstration, and commercial application budget for each fiscal year from a variety of internal offices. These funds are matched with funds from private partners to promote energy technologies for commercial purposes.

The program today continues the mission of increasing the number of energy technologies developed at DOE's national labs that graduate to commercial development and achieve commercial impact. In addition, the TCF successfully continues the DOE's technology transition "push to market" strategy with a forward-looking and competitive approach to lab and private industry partnerships.

Reflecting on the importance of delivering on the fund's mission, Drew said, "Having US taxpayer dollars focused on renewables research for us meant that we had to ensure two things. One, the research would be done in the labs and two, the research would get out into the market."

All the while the team worked together, they were in agreement that they were not going to own the term "commercialization," or claim credit for commercialization of the renewable energy industry. "We were trying to facilitate the provision of private capital to bring alive technology that was sitting on the shelf. We wanted to figure out how to find and match the entrepreneurs and strategic partners for all of that to happen," said Drew.

The focused dedication that fueled the early days of the commercialization team continues to reverberate, as each team member has gone on to conquer a portion of the corporate sector, keeping true to their experiences at DOE.

Andy had instilled in the rest of the team that their time of service would be short, but it would be successful if, "We worked ourselves out of a job," Holland recalled.

As the majority of the surviving original commercialization portfolio lies now with the Department-wide Office of Technology Transition, the team can hang their hats on success and relish the overall impact they have made.

CHAPTER 14

Be the First

Claire Broido   Mark   Loretta
Johnson   Culpepper   Prencipe

"The exciting part was founding the vision, opening the markets before anyone else and having the ability to scale," starts Claire Broido Johnson as she recalls the first days of SunEdison during a group interview I had with her, Mark Culpepper, and Loretta Prencipe.

"We never took ourselves seriously, though. We just rolled up our sleeves and were always diving into some problem we had to solve," mused Jigar Shah in a separate conversation with Emily Kirsch on her podcast, *Watt It Takes*.

Today's distributed generation (DG) sector is flourishing, as many corporations decide to focus on onsite solar generation opportunities for energy savings as well as meeting greenhouse gas (GHG) and net-zero carbon goals. These players decided that investments in solar and wind generation hundreds of miles away was not enough. In addition, they needed onsite generation to demonstrate environmental stewardship to their customers and local communities.

This new demand led to the creation of what we call the DG sector in the US renewables industry. The sector came alive around 2006 with the entrance of the first solely dedicated solar DG development players.

Having come into the DG sector in 2020, I was hooked from the first day walking into SunTribe—and then C2 Energy Capital in 2021—prior to its transformation into EDPR NA Distributed Generation.

I found the allure of onsite energy solutions mesmerizing, lured into the sector's ability to immediately impact a variety of stakeholders. Ranging from public schools and local communities to corporate and industrial entities looking to start their net-zero journeys, DG is one of the most promising sectors in the US energy market.

But this wasn't always the case, with utilities for years concerned about customer erosion due to IPP new market entrance. This concern was matched with skepticism that a decentralized grid would be able to provide reliable, cost-effective energy. The DG industry needed someone to be the first to test the concept and pave the way for additional market players.

To be the first is both an exciting and daunting proposition, no matter the industry or market sector you're in. According to *Collins English Dictionary*, the adverb form of "first" takes on the definition that, "If you do something first, you do it before anyone else does, or before you do anything else." Synonyms include "novelty," "originality," and "innovation."

Being the first of—or at—something also connotes that you were considered the "best" in defining a standard or expectation around a concept, process, or business model quite simply because there was no one else to prove otherwise.

Being the first is also associated with leadership, with a leader often being someone who was the first at something or who is the foremost of something, providing a vision, model, or a cause that then attracts other to follow. Leadership blogger Bruce Lynn points out that the most important thing for a person that "is the first" is that they demonstrate a form of leadership that at the end of the day, is all about inspiring other people.

Many factors helped ignite the distributed generation sector in 2006. The newly minted Investment Tax Credit (ITC) incentive gave rise to projections by New Energy Finance and others that solar energy indeed would become competitive and profitable as the next major source of installed capacity in the US electricity sector.

Another critical factor was the creation of the Solar Renewable Energy Credit (SREC) in 2007, when states such as New Jersey switched from traditional rebate programs that had been in place to the SREC market. Solar was coming down in cost for the first time due to advancements in technology and, combined with these incentives, became price competitive in the long run.

At the same time, the renewables portfolio standard (RPS), a voluntary renewables standard mandated by states, was starting to take off from Connecticut to Colorado. One by one, you saw states opening themselves up to commercial solar development, along with individualized programs like the California Solar Initiative (CSI) program in California that launched in January 2007.

The program was part of the State of California's launch of the Go Solar California campaign, an unprecedented

3.3-billion-dollar ratepayer-funded effort that aimed to install three thousand MW of new grid-connected solar over the next decade and transform the market for solar energy by dramatically reducing the cost of solar.

This in turn helped ignite the behind-the-meter (BTM) solar market, enabling initial solar developers who were considering going the utility route to instead focus on the distributed generation market. Seed funding and a viable business model were all that was needed.

In early 2003, Jigar Shah and Claire Broido Johnson gathered around a kitchen table to build out the concept that Jigar had begun during his MBA class a few years earlier. The plan was to take the traditional IPP power purchase agreement (PPA) model and apply it to the commercial sector.

They both had decided to quit their jobs—Jigar having previously been with BP Solar and Claire with Constellation Energy—to fully dedicate themselves to a promising gap they had identified in the renewables sector and to inaugurate a specialized onsite solar development company.

Jigar Shah started his journey into renewables knowing he wanted to be in renewables. "Back then we were told that in order to get into renewables, you had to have an engineering degree," Jigar told Powerhouse CEO Emily Kirsch on *Watt It Takes*. After college he had an internship in 1995 with an electrical engineer turned CEO named Allen Barnett at Astropower.

While there, he met Howard Wenger and many of those who later would regroup at SunPower, the American solar manufacturer and installer. After completing his engineering degree, he had held positions at a wind company in Vermont as well as with Energetics as a consultant to the US Department of Energy in Washington.

In 2000, he landed at BP Solar, which had just bought the remaining minority share from Enron in Solarex and had become the largest polycrystalline solar manufacturer at that time.

While there, he completed his MBA at night at the University of Maryland Robert H. Smith School of Business. Being at the largest solar manufacturer had its perks, and Jigar was exposed to cutting-edge technology pitches, as well as revolutionary business models that would set the trajectory for the solar business. His time at BP Solar came to an end following the dot-com boom and bust.

A restructuring at BP that laid off several team members gave him an opportunity to exit on good terms. His time there had sparked a passion to revolutionize solar. Having come up with a renewables power purchase agreement (PPA) business model as part of an MBA class project, he was going to take a risk and start a company.

To many of us today, a PPA is nothing revolutionary and is implemented dozens of times daily between developers and corporate or retail entities. For those of you who are new to the renewables or energy sector, a PPA is a supply contract for the long-term sale of all power generated by the plant to one or more wholesale offtakers to mitigate the risk of selling energy output.

PPAs were developed originally in the US in the 1980s and began after the 1978 Private Utility Regulatory Policies Act (PURPA), which encouraged the construction of cogeneration plants, whose electricity could be sold to the regulated power utilities.

According to project finance experts E.R. Yescombe and Edward Farquharson, these utilities' long-term commitment under a PPA to purchase power, "Enabled the finance to be raised for the cogeneration plant, using the PPAs as security."

The entrance of the PPA into the energy market spurred the separation of what were private-sector companies involved in power generation and those involved in distribution, creating the development of independent power projects. This, in turn, led to a competitive market where independent power producers would compete against traditional utilities in the energy market.

What makes PPAs attractive to offtakers—or the customers buying the energy produced onsite—is that the IPP or developer takes on the full risk of financing, building, and operating the power facility, committing to deliver electricity at a competitively priced rate that demonstrates energy cost savings over the lifetime of that facility.

This model intrigued both Jigar and Claire, and they wondered whether they could use it for solar specifically on commercial rooftops and create a new concept titled "solar-as-a-service."

Thinking back to their first pitches, Claire agreed that it, "Sounded a little wacky at first because who wants to put three thousand bolts in your roof without actually owning something on your roof." But they were going to give it a try. The kitchen table grew larger a few months later that year, with Chris Cook joining Jigar and Claire in further planning out the venture.

Chris brought with him immense regulatory knowledge that would assist SunEdison in its initial days by guiding the venture through regulatory hurdles as well as taking them through a strategic state-by-state assessment of where their commercial solar PPA model would work.

Claire lauds his impact. "He played a really important role in the early days by telling us which markets we should go after and why."

The team was looking at a number of different structures and comparing them to their PPA model. In 2003, hardly anyone was doing anything with commercial solar and those who were "sold projects at nine dollars a watt DC for big commercial projects and getting a four dollar and fifty cent a watt production tax incentive," notes Claire. "It was a chicken and egg kind of situation. Do you get the customer first or do you get the financing first, and how are you going to get the panels sourced?"

In 2004, they entered a business plan contest at Harvard Business School to test the SunEdison concept and won. Shortly after that, they brought on Brian Robertson, a Canadian entrepreneur who had turned down one of the top jobs at Yahoo to join SunEdison, as CFO.

Brian was intrigued by the complexity behind the concept of solar-as-a-service and lunged straight into risk models to figure out how they were going to hedge their risks and operationalize SunEdison.

He became the lead in negotiating financial agreements and navigating through "special situations," which is what onsite energy generation was called, given the first-time nature of these types of energy agreements.

Early commercial customer adopters began to line up, with SunEdison's first client being Whole Foods, who wanted to put solar on their store in Edgewater, New Jersey, across the river from New York City. Staples and Ikea followed suit. Things were looking up and in 2004, Claire structured the very first SREC contract, creating the first of many contracts that would fuel the SREC market in the Mid-Atlantic.

Much of their success in customer acquisition was due to PowerLight Corporation—one of the larger integrator/installers of large solar PV projects—"having done a lot of work in the sector already," Jigar recalls, referencing PowerLight's work with several retail, hotel, and commercial entities.

Even though others were in the space, to Claire, the difference SunEdison brought to the market was "our ability to scale."

Their success came from initial thought leadership demonstrated by corporate sustainability leaders among the Fortune 500. Claire recalled Mark Buckley of Staples as one of the "original" leaders willing to give the young SunEdison team a chance.

According to Claire, "He really was a visionary, and despite his assessment of us being young and potentially not knowing what we're doing, he gave us a shot, which jump-started us alongside our Whole Foods project." To Claire, the usual experience with corporates was they're not understanding the overall value of the proposition. "You could walk into a client back then and say, 'I'm going to lower your prices. I'm going to lower your energy costs. And it's not going to cost you a nickel,' and people did not believe you."

Jigar first funded SunEdison using his personal home equity line of credit, and their first round of external financing afterward came from Goldman Sachs on June 8, 2005, in the form of sixty million dollars in project financing at a 17.6 percent interest rate. Shocking as that rate may be to us today, it was what it was, "Because we were a bunch of early thirties punks trying to put together a solar company," says Claire.

They also got some very early investment interest from Jesse Fink, who was with Marshall Street Capital (today's

Mission Point Capital), along with corporate investment from David Busby, who remained a great champion of SunEdison for most of the company's history. Later in 2006, and despite competing term sheets from Vantagepoint and Riverstone Holdings, they received capital funding from Goldman Sachs in the amount of $26.1 million.

The team continued to grow, with SunEdison headquartered in Beltsville, Maryland, and their business model in 2005 had evolved to buy developers who could provide reliable development and installation services, which was one of their first major challenges.

The first developer they acquired was New Vision Technologies, set up by Brian Jacolick and Jeremy Page, which gave SunEdison a strategic foothold in the incentive-rich California market as well as direct relationships with the utilities that were still reeling from the California Energy Crisis of the early 2000s.

Growing out their foothold in the state, the team then approached Rick and Angela Lavezzo, who had created the fastest growing and largest EPC in the solar sector, California-based Team Solar. Rick and Angela came to solar in the late nineties after trying to put solar on their residential home.

"After doing the research the best quote we received was one hundred thousand dollars. For that price, I decided I could figure it out myself," Rick was quoted as saying in an interview with *Boss Magazine*.

After learning more about the industry, the couple formed Team Solar. They won a contract in 2001 for a one-million-dollar solar energy project for the Sacramento Municipal Utility District (SMUD) to deploy residential solar averaging two to four kWs on six hundred homes.

According to the Green Power Networking Institute (GENI), SMUD, which was the fifth largest public utility in the US at that time, can be credited in creating the first grid-connected solar program in the US.

In 1993, SMUD adopted a long-term PV commercialization strategy, inaugurated as the SMUD Solar Program that was aimed at accelerating the cost reduction of grid-connected utility PV applications.

SMUD's Superintendent, Jon Bertolino, led the successful deployment of solar throughout their territory and cemented the public utility's leadership in Northern California in the 2000s.

"We were hooked! We fell in love with the industry and started doing residential solar. Then some people asked if I would do commercial scale and then I was asked to do utility scale. Growth came super quick," quips Rick in his interview.

In 2005, Rick and Angela were approached by Conergy, a European company, with an offer to buy Team Solar, offering them a deal that consisted of a majority cash offer with a small equity share.

At the same time, the founders of SunEdison contacted the Lavezzos, offering a deal with some cash, but mostly equity. The Lavezzos believed in the future of the solar industry and wanted to remain in it, so they put all their chips on the table and officially merged with SunEdison in August of 2006.

That same month, Colin Murchie, who had joined SunEdison from the Solar Energy Industries Association (SEIA), was added to Chris's team, and would be integrally involved with Claire working to pass legislation around solar renewable energy credits in places like Connecticut and New Jersey.

Another addition that made an impact in SunEdison's approach to regulatory affairs came in the hiring of Joe Henri, who had previously been on the regulatory affairs team at PG&E.

Still another invaluable team member, Mark Culpepper, joined as SunEdison's first vice president of Strategic Marketing, assisting in the company's go-to-market strategy. Having previously come from Team Solar, his addition allowed SunEdison to scale, and employee count ballooned to over four hundred employees.

Having previously worked in tech before his renewables career, he felt a calling to transition into the growing solar sector during the second Gulf War. "I was working in tech, and when the war broke out, I felt I had to do something to get us off our addiction to oil, and I couldn't sit on the sidelines anymore," Mark says.

During their organic growth starting at the table in Jigar's kitchen, the SunEdison team found themselves experiencing being the first in this space of solar-as-a-service. The team often found themselves having to answer first-time questions from the financial sector, specifically around the residual value of solar.

To Mark, much of the initial questions they encountered—from both offtakers and utilities—as the first provider of solar-as-a-service were tied to data, specifically gaps in data in terms of supply demand ratios that had not yet been demonstrated by commercial solar installations since it had not been done before.

He remembers telling Jigar and Brian, "This is not a problem with power plants, and it's not a problem with financing. It's a data problem and how much you trust the initial data you're getting from the solar power plants. And if the data was good, it would tell the investors what kind of returns they were going to get out of their investments."

A story that came up in connection with this point took the team back in memories to when they had finally broken through to PG&E and had their first meeting with them about potential business together.

In reviewing the initial data, the team had produced on their commercial solar power plants, PG&E executives had told Mark and the team, "Before today, we thought you guys were putting generators on your rooftop solar power plants because your productivity is so much higher than any other solar plant we've seen. We couldn't believe that you were actually doing this legally." Mark recollects, "We actually started laughing and asked, you're kidding, right?"

Mark ultimately brought data into play at SunEdison, developing the industry's first global monitoring center, changing how SunEdison and the industry tracked and improved system performance.

They also worked toward another first to secure solar-as-a-service as an established category in the renewable energy market. SunEdison's marketing and PR team had brought on another team member, public relations guru Loretta Prencipe, who started with SunEdison as an external PR consultant.

Working alongside colleagues including Mark and a young new development lead named Kelly Speakes, they perfected the company's pitch around why customers like Whole Foods should prioritize solar to reduce costs and improve their environmental footprint. Focusing on the client and their challenges at all times, "We certainly weren't overwhelmed by the process," remembers Jigar. "We just kept working the problem until it got solved."

The ability to come into the market and scale was an advantage to SunEdison in being the first. Being the first also gets tied to the notion or status of being a "first mover," which according to *Harvard Business Review* contributors Fernando Suarez and Gianvito Lanzolla, "Allows every kind of company, that is an early entrant into an industry or product category, important competitive advantages, as well as almost insuperable head start."

In marketing strategy, this concept according to the Corporate Finance Institute, also known as "first mover advantage," associates a number of associated benefits, including the ability to be the first to establish strong brand recognition, customer loyalty, and early purchase of resources before other competitors enter the market segment.

However, unlike other industries, where first mover advantage sometimes leads to handicaps later down the road, such as perceived "free rider effects," shifts in technology or customer needs, or "lost profit" due to upfront R&D investments costs, the founders of SunEdison saw their market role differently.

They had to be open to being the first and to opening the path for other companies market by market, focusing on creating overall industry impact versus fixating on industry competition. "SunEdison got on the ground in the states that needed help to get the right legislation in place, or if the right legislation was in place, make sure it came into place," recalls Loretta. "SunEdison was always the fastest and first to understand the market the best."

Her impact on the company was profound, and before they knew it, "We were in *The New York Times* and *Wall Street Journal*," exclaims Mark.

It's widely agreed in the industry that while others focused on business in the incentive-strong market environment of California, SunEdison's critical work in lobbying regulators for solar carve-outs on the East Coast led to successfully opening those markets for the entire industry. This was due in most part to Claire's persistence to make the market work. Claire ended up raising SunEdison's first sixty million dollars in project financing.

Being the first came with a couple of requirements. "When building a company, the best-laid plans are always just a guide. It takes optimism and a persistence to overcome the hurdles presented every day," recollected Jigar.

"It was also all about having the right vision and the right skill sets assembled on the team," added Claire. Like many who had to be the firsts in their market segment in renewables, you had the visionary in Jigar, the executioners in Claire and Brian, the strategists in Chris, Colin and Jory, and the marketers in Mark and Loretta. Together, this team could ride out answering the first-ever questions from initial corporate offtakers and financiers on how to make the solar commercial PPA model stick.

At SunEdison's peak, the company was North America's largest solar energy services provider, with a market capitalization of ten billion dollars and more than seven thousand employees.

Years later, their hard work has led to the overall industry racking up over one hundred corporate renewable sourcing deals in the US alone, making up over ten gigawatts of capacity executed in 2020, up from a mere 1.5 GW in 2015, according to the World Economic Forum.

The underlying vision and the present industry's ability to execute these additional MWs can all be tied to the first conversation between Jigar and Claire around that kitchen table, back in 2003. Ultimately, the team at SunEdison will forever remain in the history books of renewables, credited with standardization of the PPA for use in the commercial solar market.

According to the team, a lot has changed from their initial days working on daily issues that were new to initial corporate customers.

"I remember us having to call Walgreens often in 2003 or 2004 when we saw data coming back stating that something was going on with their system," bemuses Claire with a smile. "We'd usually have to remind them to not forget about turning their inverter back on because we'd otherwise lose months of solar production due to someone rustling in the back of a store and flipping the switch off by mistake."

CHAPTER 15

Adaptability

Nancy Floyd

"It's what's made me, and I would have never told my investors this, but I didn't want to be in any other type of venture capital," begins Nancy.

"You know, I have this very strong altruistic streak. And that's been important to me that I follow that versus just being here to make money for my investors. But I wouldn't have done software or semiconductors just to be in the venture capital world. I wanted to do clean energy, and it's been so satisfying," she said to Gil Jenkins on a recent *Hannon Armstrong* podcast.

In the 1980s, Nancy Floyd's first foray into developing renewables was not going as planned. She found herself packing up moving boxes and heading back to the East Coast after only a few years of being out west.

Looking back, Nancy knew she was taking a risk moving to the Bay Area not knowing a soul when she had decided to join a real estate development company. The plan was to work together with partners at the company to attempt to build the first utility scale wind farm of thirty-kilowatt wind turbines.

After a few months of negotiating leases and observing how the business was run, she realized her business partners were not funding the business the way they said they would. In addition, they didn't really see eye-to-eye on business ethics.

When she moved to San Francisco in the summer of 1982, she knew the renewables industry had potential, and her passion prevented her from going down the traditional route taken by many in the venture business. The venture business's entry into renewables "was young, but it was still there," recollects Nancy. "And I thought, boy, there's an industry that really needs new technology."

Nancy was poised to be a risk-taker, albeit one who took a calculated approach to the professional risk she was undertaking early in her career. Her foresight drove her to assess the realities of market and the challenges in front of her in the emerging renewable energy industry.

She saw the potential to tap into the budding renewable energy industry and produce higher and more secure returns. The benefits were obvious to her, as she foresaw the fundraising potential associated with the industry. Alongside the attractiveness associated with the caliber of entrepreneurs rising, she saw an industry that guaranteed her future investors the returns they wanted.

The first reality she had to face was something culturally embedded in the energy industry. To be a woman and to be a venture capitalist were not a common overlap, nor one welcomed with open arms. The concept of a greater cost-benefit associated with a more inclusive venture capital firm—leading to greater profitability—was nonexistent.

Virtually no women were in the venture capital space, and this trend even continues today. According to Bloomberg, female founders secured only 2 percent of venture capital in the US in 2021, the smallest share since 2016 and a sign that efforts to diversify the famously male-dominated industry are struggling.

The second reality was that at that time in the United States, renewable energy projects were seen as expensive novelties—more like science experiments with philanthropic intentions than viable and profitable future sources of energy.

Development of renewables got you automatically associated with an environmental movement that was publicly greeted with a pat on the shoulder and the proclamation that "you're such a good person," but privately associated with being naive, not serious enough, and unprofessional.

That perception is almost hard to remember now, but only a handful of wind turbines were installed at that point, primarily in farmland areas with novel ones built in Fairmont, California, and a lone few on the Altamont Pass outside of the Bay Area. Getting those built in the first place had taken much politicking.

Getting up in front of investors and local government leaders to speak about the development of any kind of renewable technologies in the 1990s was unheard of, especially coming from a serious energy business leader. Indeed, a typical entrepreneur in Silicon Valley looking at the industry in the 1990s would have focused on more "viable" technologies in the software and retail sectors.

However, the combination of these two realities and the serious challenges associated with them only fueled Nancy's ambitions. Rather than signaling defeat, Nancy unpacked her bags on the East Coast and went straight into fundraising mode.

Fast forward a few months later, and with a successful round of funding behind her, she returned to California. Having watched the real-estate business that she was a formerly a part of floundering in the wind, she did the unthinkable. She bought her former partners out and became CEO.

Nancy's story, like many leaders in the renewable energy industry, demonstrates that in pursuing one's career, the journey is never linear. The well-known British polar explorer Sir Robert Swan, an Officer of the British Empire (OBE) known for his near-death treks across Antarctica, is also famous for coining the term "leadership on the edge."

"One's journey toward any goal is never ever straight," Sir Robert often says in his speeches. Indeed, periodic pauses, detours, roundabouts, backups, and reassessments are almost always faced by risk-takers. "However, you pick up where you last were no matter how behind you are on your mission and continue the journey in front of you to reach your goal."

This nonlinear path was not foreign to Nancy. She has survived and thrived through multiple bubbles and busts. Most other typical entrepreneurs or venture capitalists in Nancy's shoes would have otherwise shut down and opted to focus on more lucrative short-term investments in traditional energy or software related industries.

What differentiated Nancy's journey, and how did deviations in her journey lead to her success? What about our industry provided the passion she needed to continue forward? How did foresight combined with a realistic identification of profitable opportunities—and what to some would be insurmountable challenges—provide an invincibility of sorts? Was her persistence to succeed in and adapt to an ever-changing renewable energy industry unique? Is the commitment Nancy made in forging forward something that can be emulated in each one of us?

Back in late December 2004, I was getting my workstation set up, having been onboarded by Mike Eckhart and Jodie Roussell at ACORE the week before. My first project was to work with a conference partner, PennWell Corporation, and coordinate on exhibit booth sales. Mike's recommendation to get kick-started was to call twenty-five names on a list that he gave me.

The goal was to, "Interview them about renewables, their beginnings, their business models, and don't end a call without getting three introductions to others they recommend you speak to." Staring at the list of twenty-five, I picked up the phone—this was still the age of landlines at every desk—and began to dial.

Looking back, I had no clue I was calling the elite of the industry, many of whom would become mentors as well as friends. As I looked down at my list, I started with the first name, which read Nancy Floyd.

Nancy was highly recommended by Mike and was a managing director at Nth Power, based in San Francisco. She was one of the founding partners who got the venture capital firm established and set up in 1993. Having ventured into the industry as the first woman wind developer in 1982, she then set up the wind development shop NFC in 1983.

In setting up NFC, Nancy admitted that she "was not prepared to be an entrepreneur," but every career step she took built on previous knowledge and experience in the industry, proving her situational awareness and ability to adapt time and time over. She launched into her career in renewables when she was recruited from the East Coast by a few folks she knew in the real estate industry who had leased land in the Altamont Pass, east of the Bay Area.

Moving cross-country to San Francisco in 1982, Nancy began first as a regulator and negotiated a first-of-its-kind nonstandard contract with PG&E to buy power from the first wind farm being developed in the Altamont Pass. This new contractual structure evolved into a "Standard Offer Number Four," which was a favorable form of contract for wind acquisition. After her time there, she went on to fundraise a few hundred thousand dollars to start up NFC.

Working with her team, she ultimately grew out NFC's portfolio to thirty million dollars in wind farms and sold the company to a larger developer in 1985. Nancy's adaptability played a major role in the decision with her noting to me during an interview that, "I saw that the tax credits were expiring and to stay in business, I would have to be vertically integrated, meaning I would have to manufacture wind turbines. I wasn't prepared to do that."

She shifted her strategy based on her own assessment, which paid off in the long run, generating more than twenty-five times the invested capital she had originally put in.

The *Cambridge English Dictionary* defines adaptability as an ability or willingness to change in order to suit different conditions. They go on to use it in an appropriate example stating, "Adaptability is a necessary quality in an ever-changing work environment."

Psychologist F. Diane Barth adds that, "Adaptability is not a matter of ignoring your own feelings, needs, beliefs, or thoughts and pushing through no matter what. It's a process of interacting with changes."

In what many consider as one of the tenets of emotional intelligence, the ability to be able to adapt and the agility around it, allows individuals to break their comfort zones and think outside the box. When they unlock the ability to adapt, risk-takers surround themselves with people and subjects outside their area of expertise.

This, in turn, exposes them to new perspectives and ways of doings things that they would not have perceived by their own volition. According to journalist Nele Grantz, "It will also ensure that you avoid shortsightedness."

Through adaptability, risk-takers can combine courage with creativity to create aspirational forward-looking plans for innovative solutions that sidestep challenges, setbacks and—just like Nancy—continue to thrive, moving from one venture to the next.

Her ability to continuously adapt to the changing dynamics surrounding renewables investments continued after the sale of NFC. Nancy saw the difference that technology made in disrupting the landscape within the energy sector.

She saw small companies getting funded through friends and family and the venture business "was young, but it was still there."

Her vision inspired her in her post-NFC days to purchase the technology practice she had nurtured and run through the late eighties for a utility consulting firm. With a few others, Nancy went out to raise her first fund and christen the beginning of her success in later years with Nth Power.

"And let me tell you, it was not for the faint of heart," recollects Nancy. "We visited 197 investors around the world. I mean—we counted it—and nine signed up." Nancy and her team ultimately raised sixty-five million dollars for that first fund, which she noted was "small but ended up being very impactful."

After three years of developing the fund and taking no pay, her team invested in fourteen companies, which resulted in successful deployments of four IPOs and two M&A transactions.

Reflecting back, Nancy stated, "The next fundraise for the second fund was a lot easier. Still not easy, but a lot easier because it was still a very young category. The thesis behind investing in what we called new energy technologies, which later became clean technologies and now climate technologies, has really changed radically since the nineties."

The 1990s were all about utility deregulation, and utilities were branching out to look at differentiating in the market, hoping to find products and services that would get them to the next phase of their energy portfolio evolution.

Of the nine investors that Nancy and her team had backing them, eight were strategic investors, ranging from Électricité de France (EDF) and Duke Energy to the National Pension Fund in Sweden that had already experienced a deregulated utility market in Europe.

According to an academic white paper by Magali Delmas, Michael Russo, and Maria Montes-Sancho called "Deregulation and Environmental Differentiation in the Electric Utility Industry," during this period, "the familiar pattern of firms applying their resources to differential strategies took place. This process was stimulated by new freedoms that allowed them to view ratepayers not as an aggregate mass of demand but as an amalgam of distinct customer groups."

The piece goes on to further state that this period allowed an "institutional change" to occur that "reshaped competitive landscapes" in the energy market and allowed for the designation of public policies that could augment the supply of public goods, like a healthy natural environment.

Noting the adaptability needed by utilities back in those days, Nancy says they, "Saw the opportunities that were coming out of deregulation."

Gathering these investors, Nancy and her team launched their second fund in the 2000s, this time focused on energy supply. It was the advent of the funds sector investing in renewables, and generalist funds entered the market, seeing the needs of offtakers and the investor opportunities for grid advancement and grid connectivity to more modern forms of renewable energy.

The opportunities in the market were immense, ranging from investors' choice of various renewable power generation technologies as well as a variety of biofuels tied to diverse feedstocks. According to Nancy, two-thirds of funds raised went to either next generation solar or biofuels. Unbeknownst to her, that trend would continue for most of the 2000s.

Moving to most recent times, which mirror the adaptability needed to scale in the energy sector, we saw a renewables industry that began to shift and focus on customers getting engaged with the concept of energy—from its production to its consumption. Investments in the VC world moved from wind and solar projects to investment in small devices centered on residential application like thermostats and EVs.

Going from fifty million dollars in VC investments in the early nineties to peaking at one billion dollars in 1999, the VC industry then eclipsed that record again in 2009 and 2010 at eleven billion dollars before settling at a lower rate. During this period, renewables or "clean-tech" investments were the largest sector represented in VC portfolios.

"That has now led into this next wave of climate tech, which I think is terrific because climate wasn't even part of that discussion in the nineties, or really, even in the early 2000s," said Nancy in a recent Hannon Armstrong podcast in September 2021, prior to the COP 26 conference in Glasgow.

Indeed, her predicted next wave of funds commenced in the 2010s, churning out new investment and entrepreneurship from the next generation of venture capitalists—many of whom Nancy collaborated with—in everything from electric vehicle (EV) charger deployment to micro-grids and AI intelligence demand response systems.

The wave of venture capitalism was born on the backs of many venture capitalists who believed in the early renewables industry, ranging from Vinod Khosla of Khosla Ventures, Andrew Beebe of Obvious Ventures, John Doerr of Kleiner Perkins, and other notable female venture capitalists, Nancy Pfund of DBL and Gina Domanig of Emerald Technology Ventures.

In addition, investments in EVs themselves have ballooned this last decade, with the most notable IPO being that of Tesla, which was orchestrated by venture capitalist Ira Ehrenpreis, who is now with DBL Partners.

When asked about current VC trends on the recent Hannon Armstrong podcast, Nancy answers, "I wish I was founding Nth Power now instead of twenty-nine years ago because there's no turning back. It's the perfect time."

In reflecting on these trends in the VC world as well as how adaptability aided her during her initial years in renewables, Nancy notes, "My career has been the most exciting, unplanned curricula I could have imagined."

With the twists and turns the industry takes, coupled with its constantly evolving nature, "It's also a career where you can make impact and nothing's more satisfying. There are jobs for every interest, whether you're in marketing and communications, research, or engineering. This is not a dying field."

These remarks by Nancy underscore and aid in eclipsing her notoriety in the sector as the first woman venture capitalist. She is probably the most sought-after advisor for professional career development in the renewables sector.

Nancy's main point of advice to any professional is: "Messaging definitely matters." As much as investment was flowing into renewables in the mid-2000s, a number of failures, especially in next generation solar and biofuels, in many ways cast a doubt on clean-tech profitability and resulted in cooling down of VC investments in the sector.

This advice dovetails into an ancillary "nugget of wisdom" Nancy mentioned when asked about advice or feedback she recommended people reject, which is to, "Sell yourself and be larger than life." That kind of overinflation of details surrounding manufacturing of initial renewables technologies that couldn't deliver ultimately led to some of our industry's major corporate failures. At the end of the day, not all companies were able to adapt to the demands laid out by an industry that demanded rapid scale-up.

During the mid-2000s, many investments in the sector were capital intensive, focused on getting manufacturing facilities built instead of seed funding initial development, which many, including Nancy, did not think should be considered venture capital deals. The result was the impending withdrawal of many big-named funds who had moved away from traditional investments aimed at scaling and commercializing proven technologies.

Given this, Nancy recommends adapting to the times by shifting messaging on investments and focusing on making climate the market driver. "I think investors are interested in putting their money behind climate-focused funds. There's been a big surge of interest," says Nancy.

Her last piece of advice, which both propelled her to challenge her initial boundaries as well as offers us invaluable insight on how to be a more resilient renewables professional, stems back to her pre-NFC days when she first came into the industry with the real estate group that initially invested in the Altamont Pass.

During the nine months of her working with the group pulling together the deal with PG&E, Nancy had an awakening of sorts. She followed her own advice to "Prioritize humility, be respectful and look for the best in everyone" as she packed up her bags and headed back home to the East Coast. In following months, she would fundraise to start NFC and return to the West Coast to continue her career in renewables.

"And there I was CEO of a company not even knowing how to budget or how to forecast, and I just was a sponge and took help where I could," recollects Nancy. During this time adapting to her new reality at NFC, she made the call to invest in the first vertical axis wind turbines. Speaking with experts in the market and listening to colleagues, she understood that she was trekking into unknown territory.

But Nancy's ability to adapt enabled her as a leader to be able to detect and identify upcoming challenges and ultimately gather the knowledge or tools to ram through them or circumvent them all together. Maggie Wood Glasser, a scholar from Washington University adds that, "It begins with your ability to signal change in your environment. You must then decode and interpret these signals and ultimately act with purpose to create a positive outcome. You will unavoidably and inevitably experience successes and failures as a result of your actions. Most importantly, you will build resiliency, which by definition is the ability to recover from or adjust easily to change."

Being an adventurer, Nancy was "all hands" her initial days of working on projects in the California foothills. "I was literally out in the hills with rattlesnake guards that my farmers gave me because I didn't realize rattlesnakes were out there, nor did I realize, as I was digging anemometer controls out of the ground that there were scorpions. In addition, I didn't realize everything you needed to know going before the Planning Commissions to get wind turbines permitted."

If that wasn't enough, Nancy recalls that the commission additionally had a fear that turbines along major freeway would distract drivers. "It was the real cowboy days," she laughs.

Her decision to start anew and begin NFC—which provided the foundation for her ultimate success in the industry and in founding Nth Power—in many ways is a lesson that in beginning a career journey, not all routes are linear and spaced out with equidistant milestones or benchmarks. At times you have to divert or dodge roadblocks. Sometimes you might even have to pause. But you pick up where you stopped, adapt to where you find yourself, and continue your journey.

Nancy's journey was defined by the last piece of advice she offers up, which compares skiing to the venture capital business and assessing risk.

"It's a relationship business and one where you have to feel your way through the situation. Just like in skiing, it's a lot about feel, making decisions on the fly, and you may be literally on the fly." The elements she attributes most closely applying to venture include the ability to pivot and the ability to make quick decisions. Quite simply, it's all about adaptability.

Nancy concludes, "I mean, you're on the verge of being out of control if you're going to be good, and you win by hundreds of a second. It's a pretty amazing sport."

CHAPTER 16

Fearlessness

Rich
Dovere

Candice
Michalowicz

"Renewables sounded interesting, and I was willing to give it a few months to try it out," Candice Michalowicz remembers saying as she brainstormed with an old friend, Rich Dovere, back in 2006. "It was the right time and the right moment for me. But if Rich and I had not crossed paths at that particular juncture, would I be in renewables today? I just don't know."

It has been over a year and a half since the sale of their portfolio—C2 Omega—to EDP Renewables North America in 2021. Their journey was not linear and relied on an evolving understanding of the potential the distributed generation sector had in the larger energy market.

"The concept of having distributed investments, particularly in the power generation space, is new," opens up Rich in describing the sector to me. "So, I think that the biggest challenge to those entering distributed generation is adapting to a risk-adjusted distributed approach to development. You have to have a nontraditional project finance mentality combined with a sense of fearlessness."

I was quite fortunate to have met Rich years ago in the late 2000s in Washington, DC, when he had organized a student policy seminar on energy while wrapping up his studies at the University of Wisconsin-Madison. He'd reached out to me at ACORE, asking for me to join the seminar to discuss the intersection of renewables financing and federal policies.

Being a believer in always trying to help student groups when I could, I agreed and found myself surrounded by smart young leaders who were interested in renewables and figuring out how to catapult themselves into this new bourgeoning industry.

Rich began our conversation in a fearless way, providing a "Jeopardy lightening round" of questions that truly challenged me in my knowledge of renewables. The resulting conversation with the students that day provided thought-provoking approaches to state-level renewables policies.

That seminar all those years ago made me commit in my mind to accept at least one to two student conferences a year to keep myself updated on my knowledge of renewables policies. Looking back, little did I know that two decades later in 2021, he would call me up to offer me a job on an exciting new venture that would take me down to Texas.

When one googles "fearlessness," no clear definition shows up in the first twenty search results. Disappointed, I went ahead to define the lack of fear by associating it synonymously with a few other traits that work in tandem with it. As no surprise to me, having adventurous spirit, a lack of hesitation, motivation, inquisitiveness, and being intrepid or audacious were associated traits brought up by entrepreneurs like Rich and Candice.

Post University President Don Mroz in a piece for *Wired Magazine* reflected that when wanting to invigorate innovation, experimentation and failure are contributors to innovation. In fact, failure can be just as much of a learning experience as success, as long as one is open to the prospect and unafraid to fail.

He attributes sparking innovation in a work environment to five key factors: (1) creating a learning environment, (2) ability to adjust an attitude or approach, (3) creating a culture of trust, (4) celebrating wins and (5) experimenting as much as possible.

Having these factors in place requires a lack of fear—a rejection of the fear that would normally hold an entrepreneur back from learning or being adaptable in business approach or concept.

The same trait allows trust to be freer flowing as well as causing a "death to ego" that allows celebrating wins by team members and opens up an environment filled with experimentation of ideas and pushing of the envelope on business models. All one needs to do is to have confidence and not be afraid of failure.

Just like the coalition of the willing and the various oak trees we have heard about before, early distributed generation (DG) renewable entrepreneurs, despite being competitors, also were closely tied together. Connections ranged from individuals having worked together in previous careers or companies as well as entrepreneurs—sometimes in the same sector—being funded by the same investors.

While SunEdison can be credited as the first entity to have standardized the PPA, it indeed found itself surrounded shortly thereafter by a number of other entities that were funded and charged to break open the opportunities in the distributed generation—or as some call it—onsite energy market. Little do many know that these companies all have one common source of origination due to funding from mentors, friends, family, and benefactors—including Jigar Shah.

The DG sector could claim its origins to the year 2006, when a number of entities opened their doors. New Jersey-based Nautilus Solar Energy, started by Laura Stern and Jim Rice in December of 2006, began their venture sketching out their business model on the back of a napkin at Starbucks.

Writing notes on napkins was a thing apparently, and at the same time down in Washington, DC, childhood friends Rich Dovere and Candice Michalowicz met up at Proof, a wine bar across the street from what was then known as the Verizon Center. Sketching out their thoughts on the back of a napkin, they conversed about the new solar industry. Their initial thoughts were to look specifically at special situation projects associated with riskier solar installations on commercial properties.

Though they were both new to renewables, these young entrepreneurs had a shared trait—a lack of fear of jumping into new ventures.

Rich had come off three failed entrepreneurial ventures but was just getting started and saw opportunity in distributed generation.

Candice was wrapping up foreign language policy work with the Department of Education and George Washington University and was looking to escape DC. In their minds, there was no better time than the present to get started because they had nothing to lose.

Interestingly, being the first doesn't always start with a confident light-bulb moment and a definitive way forward, as felt by Rich. "At that time, I didn't feel like there were actually that many other options career-wise. It wasn't like I could have gotten a job in banking or gone to law school. Those doors were not necessarily open to me, so we had to open other doors."

Candice initially joined Rich in a venture he had created called Adamas Energy Investments, and though they tested a few approaches, they found themselves not getting traction and floundering in the first initial months. However, the mentorship of Jigar Shah supported them.

Jigar gave his time and connections to Rich and Candice, introducing them to industry movers and potential future funders. "Then one day, sight unforeseen, we got a wire of a million dollars in our bank account along with a corresponding note with a simple message: Go buy something," recalls Rich with a smile.

With that seed capital, they purchased a DG solar project in the "special situation" category and the rest was history. They both attribute their success to not only going into the venture without fear of what was ahead of them but also knowing their respective strengths in running the operation.

In reflecting on the daily delegation of tasks, Candice remarks, "It was a natural split in responsibilities. I really didn't want to deal with the financial model, and due to my analytical background of being in art history and understanding things in context, I was fascinated with focusing on the devil in the details."

Rich in turn preferred the financial structure of the deals and was a natural in external reaching activities, drumming up business and exercising the ability to be able to answer any questions potential customers threw at him.

DG is not for the faint of heart. It's a different beast compared to the utility scale side of the business, where your focus is on one sole four hundred MW solar or wind farm. Instead, you focus on individual onsite energy projects averaging one MW that then add up to a portfolio of one hundred MWs and up. The development process is quicker and you're constantly originating to feed your pipeline.

Summing up the main disconnect for new DG stakeholders, Rich observes that, "The thing that is ultimately the most difficult in distributed generation is understanding that yes, you're investing in an asset, but that asset is really not a standalone. It's part of an operating company."

To Rich and Candice, as well as Laura and Jim, they had an advantage compared to those who had been in the power generation sector for decades quite simply because the naysayers of decentralized energy did not bog them down.

They didn't have DG deal expertise or market knowledge as a deterrent to hold them back or scare them off. They started their corporate business models from scratch, approaching the market in their own way with no previous bias impeding their strategies on how to scale up, acquire industry data, garner client relationships, or generate profits.

These entrepreneurs were not fearful of the variability of a sector that had them focusing on a one MW project on a Tuesday and then developing a three MW project in the same client portfolio the following week, and so on.

The approach of focusing on reaching the ultimate goal of one hundred MW per client was the main driver of their ambitions, achievable only through development of deep relationships, proven success, and adapting to the mentality that, "not one shoe fits all" in DG.

This new mentality or way of looking at the grid in a distributed way isn't easy for many. "So, a lot of professionals have been doing it the same way for over one hundred years, and in the last eight to ten years, you're telling them this new way has a lot of the same way of thinking, but we've got to do it differently," adds Rich.

A DG entrepreneur has to have a different mentality due to projects being situation specific, whether the type of installation needing to be customized, the geographic location of the installation, or the specific client—be it a Walmart supercenter, a Facebook data center, or a healthcare facility run by Scripps Health.

In addition, the importance of energy reliability and security associated with the grid have come into play here in the US. With the advent of weather cycles influenced by climate change as well as more frequent external threats of power system hacks, onsite sources of energy are in ever-greater demand.

DG in many ways has revolutionized the concept of self-determination and self-reliance as it applies to energy choice, energy access, and onsite energy generation. "This makes it a very American thing," beams Rich. As a US History and Political Science major in undergrad, Rich links the way America likes to export American democracy and values to the way the US industry approached the distributed nature of this budding sector of renewables.

"A customer, be it a corporate or a public school, when posed with the freedom of choice—to be tied to a utility and its rate schedule, build a solar plant on their facilities, or maybe even negotiate their own contract—will err on the side of choice and independence," he remarks.

This in turn makes DG increasingly attractive to customers and very powerful. Rich professes, "Once that customer goes distributed, it's unlikely they're ever going to go back to being solely dependent on a centralized grid. That moment of realization—both for the developer and for the customer—is a moment to ponder upon. The minute you switch on that onsite solar or storage facility, you have created a commercial break from a 150-year power generation model, leaping into the future of power generation."

Throughout their journey in growing Adamas and C2 to EDPR NA DG, Rich and Candice lacked a sense of fear that would have paralyzed them in their decision-making.

According to psychotherapist and professor Dr. Theo Tsaousides, fear in certain instances is a good thing that dictates certain actions that can steer you toward a positive outcome. One has only to reflect on the "fight or flight" type of fear, which supposedly exists as a genetic instinct from our ancestors and can be attributed to their survival in the wilderness.

He goes on to claim in his 2015 *Psychology Today* piece "7 Things You Need to Know about Fear" that fear can be as much an ally as it can be an enemy in that the more real the threat, the more heroic your actions.

Though I do agree with him, I believe that fearlessness—including the absence of "positive" or "motivational" fear—plays an equally and uniquely important role in renewables entrepreneurs as does fear in other entrepreneurs who either have fear and successfully navigate through proactive action to a desirable income or fear that sabotages their chances.

Susan Peppercorn in her *Harvard Business Review* piece on "How to Overcome Your Fear of Failure" notes four strategies or characteristics implemented by those who lack fear in approaching some of their hardest challenges.

To Peppercorn, it all starts with redefining what failure means to you, specifically evaluating the discrepancy between what you hope to achieve and what you might achieve, helping you focus on what you've learned. By doing so, you begin to recalibrate future challenges and your approach to them by mitigating or eliminating fear.

Another strategy useful in approaching situations without a lack of fear is setting "approach goals" versus "avoidance goals" and focusing on what you want to achieve versus what you want to avoid, understanding the "cost of inaction" associated with avoidance goals. Lastly, a focus on learning when approaching any situation prepares and wires the mind to garner value no matter the outcome from any situation.

As a marketer who is a productive narcissist, I know failure is inevitable. From not aligning messaging correctly to a customer audience to not selecting the right product to engage a customer—missteps happen. In my early career at ACORE, I personalized the failures I encountered and developed "avoidance goals" to avert projects that inevitably would have sharpened my skill sets. This was all due to fear and not letting myself become vulnerable, due to the longing to avoid chastisement by colleagues in the industry.

For all of us in the marketing profession, image is everything. When I grew more confident, I realized one's personal brand has to include stretch goals to demonstrate leadership among your peers.

In order to approach these stretch goals, you have to take fear and flip it upside down to make it into courage and persist in your ambitions. That's required of a risk-taker.

Reflecting on what empowered them to launch into their ventures without a lack of fear, Candice responds, "We were very fortunate to be surrounded in this industry by people who supported my growth and made sure I had a voice and seat at the table."

According to Candice, we still have much work left to be done in enforcing the importance of diversity in both socio-economic and racial background and diversity in innovative approach to the DG renewables sector.

In many ways, new entrants need to be able to see those like themselves in renewables in order to have the courage to lunge into the industry. Candice concurs stating, "It's very important for the next generation to be able to see and to have the ability to forge a trajectory for growth in an industry that requires flexibility and determination without fearing being held back."

CHAPTER 17

Innovation

Dan Shugar

"In regard to innovation, you must encourage people to take risks and try new things. We have an ethos at the company where folks know you've got their back. If somebody tries something new and it doesn't work, that's okay. Additionally, allowing organic creation of ideas from anybody within the organization is really important, as well as expanding the universe in which ideas come from—specially to include customers," remarks Dan Shugar in an interview with *Solar Power World* in 2019.

On the other side of the US from Washington, DC, and going back a decade from the DG revolution we just read about, an innovative entrepreneur named Dan Shugar was on the cusp of revolutionizing the business around solar panel racking and mounting.

"Well, I got into solar by pure dumb luck actually," responds Dan on his start in the solar energy industry. He had started his career in June of 1986 working at PG&E, the northern Californian utility, doing work in the traditional transmission and distribution, planning, and operations.

At that time, PG&E was the leading utility research company focusing on innovation behind the burgeoning renewable energy industry. When he was asked if he'd be interested in a rotational assignment with the renewables R&D team, he was intrigued and accepted. Little did he know that his decision would forge the direction of his career.

In my mind, Dan had always been associated with pioneering and cutting-edge solar technology. I first met him in my twenties while at my first Solar Power International (SPI) conference hosted by the Solar Energy Industries Association (SEIA) and what was known as the Solar Electric Producers Association (SEPA).

Dan had an uncanny ability through humility, passion, and a "rising tide raises all boats" attitude to inspire others to innovate, no matter their technology pedigree in renewables nor their tenure in renewables.

Being a serial entrepreneur, Dan was undeterred by the challenges we commonly saw in the early solar photovoltaic sector. From challenges surrounding initial physics with advanced solar design and facing the stark economic realities of the early 1990s to competing in a tough regulatory environment, Dan remained undeterred.

Dan taught me to look beyond all that and have confidence in the power of the technology we were creating in the US as a solution to the growing number of climate issues around us.

One of the other key lessons Dan taught me was to focus my energy and innovation on doing one thing well—multiplying your success with that one thing over and over.

Through his reincarnations in the solar sector, Dan has demonstrated that time and time over.

The term innovation has been severely overutilized when it applies to any industry based on technology. In fact, many will go as far as to say it's a cliché term. A 2012 *Harvard Business Review* study entitled "Innovation Is a Discipline, Not a Cliché" defended the need for us to keep including the word innovation as a necessary value sought after in both individuals as well as corporate mission statements. But the key was to disentangle a lot of associated beliefs or concepts that over the years were added to bog down the trait.

The piece argues that three misconceptions are associated with the word innovation. The first is that people get the concept of creativity and innovation confused. Even though creativity is needed in formulating an initial idea, innovation takes hold throughout the process of launching a product with the ultimate goal of creating an impact.

Innovation to the authors goes beyond discovering an opportunity and into "blueprinting an idea to seize that opportunity and implementing that idea to achieve results."

The second misconception is that only a select few—be it creative people, marketers or folks in lab coats—should drive a company's innovative activities. This is the ultimate death of innovation. Everyone in an organization should be involved in innovation in order to bring diversity in approach to ensuring ultimate and successful impact around whatever is being innovated.

The third misconception surrounding the word innovation is that is has to be associated with a "big bang." Not all innovation is earth-shattering, but being able to reliably advance the ball further down the field versus making the end goal with one kick is key in being innovative. "Pushing for big bangs often leads to overly risky ideas that have little hope of getting approved at most companies," cites the article.

In many ways, Dan's piece of advice to me on "not trying to master it all but being good at one thing" in many ways ties to this last concept of innovation as a persistent and steady advance that may seem slow now but over time leads to a greater outcome. This slow but steady innovative approach brought Dan face-to-face with his first venture.

Having been exposed to the solar industry during his PG&E days, Dan moved on to become VP of Sales and Operation at APS and NWP Corporations in 1993, which manufactured and built some the world's largest thin film systems as well as developing some of the first innovative tracker technology. While he assisted in APS's sale to BP Solar in 1995, he was approached by Tom Dinwoodie, who had just invented the first lightweight penetration-less solar roof system designed for a low slope or flat roof.

The team tested that system on a test facility and gave Tom some affirmative feedback, which in turn gave him the premise to create a company named PowerLight Corporation. A few months later, Tom reached out to Dan and asked him to join the company.

Dan remembers the conversation he had with his wife at the kitchen table that evening after he received the offer. At that point, Tom Dinwoodie was two years into his venture and had very little funding to back his start-up, operating out of a one-car garage in the Bay Area of California.

"How little take-home money can we afford in order to not have things go super bad," asked Tom to his wife. She gave him a figure, and after negotiating with Tom, he got the bare bones salary he needed to join Tom at PowerLight as president in 1996. Dan believed in the innovation that would come out of PowerLight and later fearlessly used his modest home to serve as a line of credit for the company.

They applied for, and received, a number of R&D grants from the California Energy Commission, the US Department of Energy, the National Renewable Energy Lab, and New York State Energy Development, Research and Development Authority, in essence becoming what Dan referred to as the "poster child" of how R&D seed investments, when applied strategically on what was needed in the market, can really work out.

These initial investments helped the company launch into their first few years of selling racking commercially with one of their first big projects in 1996 being tied to the first financed PV commercial system in the country. Of the projects they initially launched into, he recollected the innovation that went into helping get their Powerguard product scaled up, through a series of projects they did for a customer in Hawaii.

Hawaii, with the encouragement of HECO—the Hawaiian Electric Company—under the leadership of Art Seki and others, was looking to roll out solar across the state. They advocated to make Hawaii attractive for solar investment, complementing the already amazing amount of sunny production days they get with a 35 percent state solar tax credit.

Originally enacted in 1976, the Hawaii Energy Tax Credits allowed individuals or corporations to claim an income tax credit of 20 percent of the cost of equipment and installation of a wind system and 35 percent of the cost of equipment and installation of a solar thermal or photovoltaic (PV) system.

Dan was instrumental in structuring one of the first operating lease models and procuring financing in hand for that initial project with the Mauna Lani Hotel on the Big Island. What made the project that much more significant was that they were—for the first time—not asking the client to write a check to commence construction.

PowerLight did its own financial analysis and worked with the lessor, who in turn collaborated with PowerLight to monetize the Hawaiian Tax Credit, the existing federal Investment Tax Credit, and what at that time was a period of accelerated depreciation.

In combination, those mechanisms provided value worth over half the cost of the system over the first five years. The customer then would pay a lease rate over the term, which could range from seven to ten or twelve years. The sum of the lease payments was less than the capital positive view if you didn't have the tax credits involved. In essence, PowerLight was able to develop a financial model where the system provided positive cashflow from the beginning.

PowerLight, with its lean team of eight employees, ended up contracting a series of six projects for the Mauna Lani Hotel, including a one-hundred-kW project, which was "really big for commercial installation at that time," asserted Dan.

As Tom and Dan grew out their customer network that focused on rooftop installations, they quickly found that their customer had much higher energy demand than what could be supplied by behind-the-meter (BTM) rooftop solar, so they launched into commercializing single-axis trackers at scale in the early 1990s, based on an IP that was already out there. Later they developed low-cost controllers and other critical components needed to scale the horizontal tracker industry.

Filled with a sense of innovation, Tom and Dan had aspirations for the company going global, and in 2004 they embarked on building the world's first ten megawatt system in Bavaria, Germany. Prior to that, the largest system they were involved with was a four megawatt system in Arizona.

Through the project, they learned the ropes, acquiring development rights, securing financing, and getting their tracker certified in Europe. It was also the first time innovating and implementing a "solar inverter in a container" concept that was used in a utility plant of that scale, and to their surprise, they were able to get it all done in the span of a year, which was quick for that time.

"We create that amount of material before breakfast each day now." Dan smiled. He went on to say that their success resulted in sustained 85 percent annual compounded growth for PowerLight for the foreseeable future. The team at Power-Light continued to grow, as did their offering in the market. Dan noted the importance of identifying passionate people to drive innovation once you've made a decision to embark on a venture.

In recollecting those days, Dan leaned back and with a laugh recalled an expression a friend of his once used. "Some people read the news and others make the news. Finding people who want to make news and indeed create it—be it solar products, services or business models—is vital to any business venture."

The business thrived, balancing commercial and residential "resi" installations, and they developed a resi roof integrated system called "Suntile" that for several years had around 60 percent of market share. The company also launched into providing carports solutions.

Just as things were cruising, 2005 came along with an acute shortage of raw materials for solar, including acquisition of solar panels. This resulted in large solar manufacturers, like

PowerLight's main supplier Sharp, trying to control where PowerLight could ship panels, given constraints on different solar products they were supplying to different markets with different certifications.

Challenge sparked innovation—again another critical trait associated with a lack of fear—and PowerLight decided to start making its own solar panels in 2005. PowerLight grew to be the market-leading installer of large, multi-hundred-kilowatt commercial rooftop and ground-mounted solar power plants in California, New Jersey, Nevada, and Hawaii.

The company also grew to provide complete resi solar system solutions to more than a dozen leading production homebuilders in California. In Germany, Spain, Portugal, Italy, and Korea, PowerLight designed, developed, operated, and maintained solar electric power plants ranging from one megawatt to more than ten megawatts, including two of the world's largest solar electric power plants.

As of 2006, over ten years, they had deployed hundreds of large-scale solar systems with a total capacity of more than one hundred megawatts. The market for commercial solar had expanded globally, and PowerLight saw competition from newly created competitors such as Conergy in Europe as well as BP Solar—who was in the process of acquiring Solarex—and SunEdison in the US.

Tom, Dan, and the team wanted to take the company public and drafted an SEC S1, which provides a general overview of a public offering to market, to raise equity needed to deploy into the next phase of PowerLight's evolutions. They were approached by SunPower Corporation, the Bay Area-based manufacturer of the world's highest-efficiency, commercially available solar cells and solar panels, led by Tom Werner.

SunPower offered up an opportunity to do a classic vertical integration model. Both companies were making revenue in the neighborhood of $220-240 million with no strategic overlap between the companies, providing a good complementary fit.

In November of that year, PowerLight was acquired by SunPower Corporation. Dan went on to be president of SunPower Systems Corporation, which provided two-thirds of revenue for the whole company. He went on with his team at SunPower to take advantage of the competitive edge a combined PowerLight and SunPower solar offering would bring to market, helping conceive and design solar panels that would be optimal for trackers.

With the rise of silicon costs and the later economic crisis, SunPower Systems Corporation navigated around issues faced by competitors given Dan's ability to maintain supplier relationships. Even though they were part of SunPower, Dan was able to maintain legacy supply agreements with seven suppliers including Sanyo, Evergreen Solar, and Q-Cells, basically acting as a systems channel—a supplier model not widely adopted by the solar developer sector—giving SunPower yet another competitive edge.

What we saw in Dan, as well as his team, was that with greater outflow of innovation also comes a lot of hard work. In a *Scientific American* piece published in 2020, University of Chicago behavioral scientist Oleg Urminsky, explained why one works harder when you sense yourself getting close to achieving a goal.

According to him, there is a concept called the "goal gradient hypothesis" that states that, "The closer we get to completing a goal, the more motivated we are to continue working on it and achieve that goal." Innovation in much the same way is like that endorphin needed to spur oneself to overcome a challenge or to create a solution to a challenge to get closer to an ultimate goal.

This "progress illusion" as Urminsky calls it, suggests that maybe the actual distance to the goal doesn't matter as much as "our perception of that distance." In many ways, ultimately the goal of the leader or the team driving innovation is to hold constant the objective distance to the impact they are trying to achieve, utilizing innovation as a bridge between the opportunity and ultimate success.

True to his innovative ways, Dan reinvented himself in 2013 by establishing Nextracker when he and his cofounders decided to set up a company that would innovate, redesign, engineer, and build out an innovative system from the ground up after not finding solar trackers that met their standards.

This all came about when a customer at Dan's request had purchased a short row of solar trackers from a substantial company in the market that didn't work well and were hard to install. The trackers came with a confusing installation guide, and no one answered the dedicated customer service line. That frustrated Dan, who has a customer-first mentality. He felt he knew the space and thought they could reimagine the solar tracker differently with a developer and owner mentality for the thirty-year life of the plant.

"We were undercapitalized when we initially launched the company, and the biggest risk return was developing a new electronic controller, which is no joke," admits Dan. "We literally made a bet on technology and innovation and were essentially able to go from zero to one hundred as a leading provider of trackers after two years of operation."

Reflecting back on the theme of innovation in conjunction with joining PowerLight and later growing out SunPower and founding Nextracker, Dan credits his ability to think objectively and not get wedded to one path as a major driver in his success. At the same time, innovation partnered with passion can achieve an objective.

"Part of passion is being relentless and not taking no for an answer." Spoken like a true innovator Dan said, "You need to keep being forward-thinking about how you're going to drive results over the finish line."

CHAPTER 18

Enterprising

Kelly Speakes   Jeff
Backman   Bishop

"I realized I needed to get a real job," Kelly started with a laugh. Kelly Speakes was a young student in the 1980s, studying mechanical engineering at Boston University. While waiting tables on the side, she realized a broader energy journey was out there waiting for her.

At the urging of her mother to come back home to Ohio after graduation in 1990, Kelly got a job with an HVAC design engineer firm and was there for four years.

She realized the job she had taken was just not for her and lacked a culture of enterprise and the adventure she sought in starting a lifelong career. Kelly began to canvas opportunities with those closest to her and got in touch with one of her fellow electrical engineering college roommates to brainstorm opportunities in the market in 1995.

Kelly's former roommate mentioned that she worked at a company that was placing an order for engines for cogeneration plants. "But it was all dependent on those suppliers having a US subsidiary," recollects Kelly.

The opportunity to set up a subsidiary for a cutting-edge renewables company intrigued Kelly's roommate. In her conversation with Kelly, she asked whether Kelly was willing to join her in setting up the venture. Knowing that she didn't want to do any more design engineering and sensing the ability to flex her enterprise skills, Kelly said, "Yes, let's do it!"

With that yes, she and her roommate embarked on setting up the US subsidiary of an enterprising company named Jenbacher, relocating back to the metro Boston area in Norwood, Massachusetts.

To be enterprising is a key skill set in setting up any venture in renewables. However, sometimes enterprising entrepreneurs get unfairly pegged as being opportunists. I'd like to claim there is an immense difference.

According to University of Wisconsin Professor Paul Brians in his book *Common Errors in English Usage*, the label "opportunist" usually has negative connotations. "It implies that the people so labeled take unprincipled, unfair advantage of opportunities for selfish ends. Opportunistic people are often also regarded as exploitative. The term is often used to label unscrupulous politicians who seek to manipulate voters in their favor by exploiting certain issues or opportunities in an unethical way."

Unlike opportunists, enterprising entrepreneurs like Kelly and her roommate take, "Legitimate and skilled advantage of opportunities that spring up," according to Professor Brians. Utilizing enterprise skills sets, enterprising leaders can rally support around entrepreneurial ventures, developing viable business models built on calculated risk. Also, unlike opportunists, enterprising leaders are decisive in their decision-making—setting a way forward and sticking with their decisions.

Jenbacher's profiles and its ability to scale in the US market was exactly what Kelly needed to discover her passion for enterprise. Jenbacher, which began in 1957 and is still based in the Tyrol region of an Austrian in a town called Jenbach, went from being a small company to growing to over eight hundred employees when Kelly was with them.

It specialized—and continues to specialize—in lean burn gas engines, including cogeneration plants and power generator sets that it had been producing since its beginning.

Jenbacher engines ran on natural gas, landfill gas, sewage gas, biogas, mine gas, coal gas, syngas, and now on hydrogen. The company was acquired by GE in 2003. What distinguished the engines was that they were being utilized for the first time in industrial applications where gas would normally be flared off or released into the atmosphere, turning what was considered waste into energy.

"They also worked on rapeseed oil, so they were super-efficient, high-end, reciprocating engines for physical plants," mentions Kelly. "It was my first encounter with renewable energy and being part of a company with innovation in its DNA was just super cool."

Wanting to learn from the ground up to garner a better understanding of the enterprise, Kelly started first in project design and then moved into sales. Joining a lean sales team composed of her friend and another Austrian colleague, she relished the true enterprising start-up feel of the US subsidiary.

"And ever since then, I just knew I wanted to do all kinds of things tied to getting a start-up structured and going." Together, the team grew the company with this innovative technology, driven by a mission to decarbonize and decentralize the energy matrix of the future.

What fascinated Kelly was that you could take wastewater treatment plant gas, adjust the various ways you can use that energy that's already existed, and make power. Ultimately, she professes, "You save yourself money, do it more efficiently, and save the planet."

However, the case had to be made and Kelly admits, "Sustainability wasn't a big thing back then, so constantly being challenged by how we could convert this waste energy drove me."

An enterprising spirit became a constant theme in Kelly's life, moving on from Jenbacher to Finnish engine company Wärtsilä. Later, she worked with fuel cells at United Technologies, trying to innovate how buildings could consume zero net energy.

As is true with the renewables industry, the ability to be enterprising and scope out the opportunities in the market brought together Kelly and another risk-taker we know— Jigar Shah.

Joining SunEdison as a marketing director, she was part of efforts to remove policy barriers being strategized and looking specifically at renewable portfolio standards (RPS) as drivers to renewable energy adoption.

"We were all about removing barriers," professes Kelly. "Thinking about some of these barriers and looking at how the RPS is paired up in the states got me thinking I wanted to get involved in state energy policy for clean energy."

Her enterprising leadership trait coupled with her now long-standing technology expertise found a new calling to drive clean energy policy. Little did she know this was the beginning of a whole new career.

In the early part of 2004, Kelly had heard that Governor O'Malley of Maryland was laying out one of his newest pieces of legislation aimed at getting Maryland to a 20 percent renewable portfolio standard by 2020.

Having been in touch with the Maryland Energy Administration through her time at SunEdison, she approached then director of the MEA Malcolm Wolff—who today serves as the president and CEO of the National Hydropower Association—and pitched him.

Sensing an enterprising opportunity to create a new market paradigm for renewables, Kelly told Malcolm, "Let me be a consultant to help you on a game plan to figure out how all these various renewables technologies will get us to the Governor's goal." Given Kelly's now-extensive technology background, this was an offer Malcolm could not decline.

After consulting Malcom and being involved in the successful implementation of Maryland's first RPS in 2004, Kelly garnered a strong leadership standing in policy circles. In October of 2010, she became the Clean Energy Director for the State.

Kelly viewed her challenges as a "Rubik's Cube with different variables, issues, and problems related to greenhouse gas reduction." She admits, "I loved experiencing all the different parts of it."

Her enterprising approach to policy led her to be appointed by Governor Martin O'Malley in 2011 to become a commissioner on the Maryland Public Service Commission, serving on the five-person commission to regulate electric, gas, water, and telecommunications public utilities as well as for-hire transportation companies doing business in Maryland.

As an adjunct, she also served as the chair of the board of directors of the Regional Greenhouse Gas Initiative (RGGI). Fellow commissioners—specifically Douglas Nazarian— helped mentor Kelly into her new role, proving that no matter what career you're at, mentorship always continues to play a critical role in success.

Kelly focused her new role at the MEA to champion the Offshore Wind Energy Act of 2013, which required a maximum of 2.5 percent of retail electricity to be generated from offshore wind starting in 2017. The act also promoted Maryland's economy by requiring offshore projects to favor in-state manufacturing to be considered by the Maryland Public Service Commission.

An ancillary side-win with this act was that it also revised the RPS goal to source 25 percent of all electricity consumed in the State from renewable energy by the year 2020 and created a carve-out for offshore wind not to exceed 2.5 percent (about five hundred MW) of the overall RPS.

After her political appointment was successfully completed, she spent time with the Alliance to Save Energy, which is where I first met Kelly in 2015, working together on policies that would unite energy efficiency—or "the first renewable energy" as then Alliance President Kateri Callahan would add—with the rest of the renewable technologies being fought for on the Hill.

After having a successful run, Kelly took some much-needed time off to reflect on her career. Then the opportunity called again to tap into her enterprising approach with technology. The board of the newly started Energy Storage Association (ESA) was looking for their first CEO to help navigate the burgeoning new storage sector and navigate it through financial and market barriers.

This was Kelly's opportunity to create an oak tree for the US storage industry and finally thrust the industry into the broader energy market. "I saw energy storage as being one of those technologies that was certainly beginning and on the uptick," reflects Kelly.

To her, the cost of storage "was not the only major issue but the ability to scale storage to address energy grid reliability and resilience were major themes that reverberated to me, given my work at the Public Service Commission." On a more serious note, she added, "It was actually our legal obligation to ensure it."

The cost of storage was coming down tremendously in 2017. However, deployment pickup—specifically in lithium-ion batteries—was nonexistent and other battery technologies were still in research.

"The idea that you could unlock the potential for solar, wind, and other intermittent resources was fascinating to me," disclosed Kelly.

She had reflected on her career—which spanned engineering, marketing, strategic planning, and business development—and found storage to be the at the center of the modern energy paradigm.

Viewing storage as the "central hub to a twenty-first century electricity grid," Kelly says it was time to, "Really step storage up to be an equivalent to some of the other clean energy technologies. There was a messaging problem around that approach."

People were thinking of storage as only a renewable technology, and that's not how storage plays into the regulatory environment. Storage is a tool for the grid to become more reliant, reliable, and resilient, and storage as a renewables technology didn't make sense to some folks.

"Technically it's not a resource or a generation source. In that policy messaging dialogue, I felt I could make an impact." Kelly jumped at the opportunity.

While at ESA, many enterprising storage leaders emerged and joined the board to build up the US storage sector. These included two consecutive chairmen of the board—Troy Miller at S&C and Chris Shelton with AES's Fluence Energy.

Kelly in many ways credits both with coaching her and her team on regulatory construct, how the value of storage could be monetized, and the overall value of bringing the business case of storage to the marketplace. Together they brought storage to the forefront of dialogue in energy circles in Washington, DC, as well as in boardrooms from Wall Street to Silicon Valley.

Another key innovator was Craig Horne, who was with Swinerton Renewable Energy and was one of the first storage technologists. Craig spearheaded the concept of the storage integration business, including figuring out approaches to solar-storage hybridization.

An additional key mentor of Kelly's on the ESA Board was Karen Butterfield, who was previously with SunPower and at that time was the Chief Commercial Officer at STEM, which was beginning to gather momentum in the storage sector in 2014, providing energy storage services sited at commercial buildings to manage their energy costs.

The last board member who came up in conversation as an enterprising leader in the storage sector was someone I've known from his last days at Brookfield Renewable and initial days setting up a venture called Key Capture Energy. That leader is Jeff Bishop.

When thinking of who made one of the largest impacts in getting storage into the market and making it work, we can't forget Jeff and his partner who cofounded the venture, Dan Fitzgerald. I remember vividly the day he and Dan sat in front of my bosses—Rob Sternthal and Nick Sangermano at CohnReznick Capital in New York—and pitched their venture as part of their initial fundraising round.

Jeff had studied computer engineering at Rice University and thought his career trajectory was leading him into that sector. His time at an internship working on wind farms in Morocco with Andy Karsner awoke his enterprising ambitions and further interest in renewables. It "just changed everything," recollects Jeff during a *Recharge by Consult* podcast interview.

When reflecting on his own conversations with others that had started in the renewables sector, he mentions that it, "Always seems like whenever I talk with people and I find out about their career progression, people kind of stumbled into the area and then became passionate about it."

Jeff had begun getting interested and seeing business opportunities in battery storage around 2015 when working at Brookfield.

"I started seeing the same cost curve for battery storage that existed ten years before for solar panels. Except I knew for battery storage that, unlike market mechanisms such as feed-in tariffs utilized in Germany and Ontario or SRECs in New Jersey, electric vehicles were going to be driving storage." In turn, this would benefit stationary storage, with utility scale storage benefiting from a rapidly declining cost curve.

To Jeff it all then, "Became a question of at what point does it make sense?"

Ultimately, what was apparent to Kelly, Jeff, and the other board members was that whenever you develop a utility scale battery storage project, it can take anywhere from two to eight years to go through the interconnection process, the permitting process, and finally the alignment of commercial contracts. "And this is all before you can build it," notes Jeff.

Defining "the point of it making sense" for the storage sector will continue to require enterprising entrepreneurs willing to take a risk. Looking forward, Jeff states that an overarching goal for the storage sector will be to, "Eventually have batteries replacing natural gas peakers—and in some instances—have solar plus storage facilities replace baseload. This is where batteries really come into play. Where we can be closer to load, we can make sure that transmission lines are always utilized by drawn-down power whenever price signals are indicating pushing them out whenever they're not."

As with many enterprising leaders, the ESA Board—and most specifically Kelly and Jeff—demonstrated innumerable common traits. These leaders are known for their commercial awareness as well as their ability to be innovative and be "original thinkers."

At the same time, they can be decisive in making decisions. Being able to prioritize as well as solve problems through strategic thinking, they host the inert ability to galvanize a team that can be adaptable and flexible to the business environment around them.

A major part of an enterprising leader's skill set stems from their ability to influence and gain agreement and support from key industry stakeholders. In essence, you must have sales or development skills to sell a concept, venture, or your vision of success. Kelly and Jeff garnered the ability to sell and influence early in their careers.

Another key part of their success was their work ethic, which was fueled by a dual combo of drive and being results oriented. Most notably, they both embodied the ability to be resilient. Relying on their enterprising spirit, they focused on market opportunities versus being deterred by setbacks experienced by early storage technologies.

Being able to structure the go-to-market strategy around how to further affect the decline in storage technologies costs was just the beginning of the adventure for Kelly and Jeff. They were able to successfully sell the importance of demand response analysis—i.e., real time data—and energy dispatchability in growing out the sector. This in turn gave them the ability to provide a vision of a more resilient and reliable grid, unlocked by the potential of micro-scale storage.

In true enterprising mode, Jeff remarks, "The key thing about this space is that you need data—and the amount of data and software that is needed is incredible."

In the "traditional" renewables sectors of solar and wind, you have intermittent electrons that you can push out to the grid whenever the sun is shining or the wind is blowing. However, with batteries, "You're charging and discharging based upon market signals. And those market signals change both in the day ahead in real-time markets," Jeff points out.

Building on this, Jeff makes another point to state that setting up enterprising ventures around AI software will be needed in this next evolutionary phase of the industry. We as a renewable energy industry will need to run all sorts of scenarios to be able to concretely figure out the capabilities of batteries and how they intersect with the regulatory market.

"We will eventually start to integrate storage in mobility, electric vehicle charging, and distributed energy resources with all of those sectors participating in the wholesale market," adds Jeff. "Data really matters."

"We all had start-up company mentalities working to get these battery-storage technologies out of the lab and into the marketplace, which was an important step and the timing was right. We were really lucky," reflects Kelly. True to her enterprising spirit, she acknowledges that, "It was the right time and the right place and the right problem and challenge out there."

In looking back at the early days, Jeff admits, "We knew where the industry was going, but we didn't have a clear path or trajectory. But that's what makes it exciting."

Enterprise was at the heart of the challenge, drawing on opportunities from a technology, market entry, and financing perspective. It again felt like a Rubik's Cube full of enterprise potential to Kelly since, "There were so many values for so many different players; you had the right problem and the right sort of attribute to appease a big stakeholder set ranging from utilities, developers, and meter storage folks to commercial entities with distributed solar on their roofs and the US Department of Energy."

Her innovative approach to solving the storage paradigm led to her being nominated by President Biden in 2021 to the post of Principal Deputy Assistant Secretary at the US Department of Energy, which is where she is today.

Kelly and board members like Jeff Bishop have the zeal for enterprising ventures and new approaches in markets, policies, and finance meant to open new opportunities for storage in the market. In turn, their enterprising trait released them from the constraints of a traditional predetermined energy career track.

Kelly's biggest piece of advice if she were to speak to her twenty-year-old self is to, "Not sweat the small stuff. I was a worrier in my early years, but in retrospect what I did was focus on a problem set—and that was how to grow out clean energy in the market."

Reflecting on whether to map a career path at the beginning of your renewable journey, she comments, "I ended up doing whatever was interesting to really solve a problem set, and that set out a much more interesting career path versus having a clear directed path."

To Jeff, the challenge continues for those entrepreneurs pushing the envelope to expand the storage industry. "The biggest hurdle that still remains is market participation. Every regional transmission, organization, and independent system operator has their own ways that resources can get access to revenue. Getting those rules worked out so batteries can participate is paramount."

When looking at the integral role storage will play in a twenty-first century energy grid and the enterprising ventures still left in front of us as an industry, Jeff states, "We have a fascinating electric network where a lot of generation will need to be pulled together. Especially when solar is far away from the load and we don't have enough transmission lines to connect."

This need to pull together all this new generation will require growing out the renewable energy industry's workforce. Asked what his concluding thoughts were about enterprising traditional oil and gas market players coming into the storage sector, Jeff maintains, "I'm okay with them coming into my space. Everybody knows power is going to be here. And that's exciting. Everybody is going to be needing to play their part for us to be able to beat climate change."

CHAPTER 19

Grit

Bill Holmberg

"Yes, my father was a tree-hugger—perhaps the toughest one ever," writes Mark Holmberg, a Virginia-based newspaper columnist, in an obituary piece for the Richmond Times-Dispatch in 2016. "He believed with every fiber of his being that integrated, sustainable agricultural energy systems are crucial to the economic and physical health of this nation and the world."

Mark's father was Lt. Colonel William "Bill" Holmberg. Bill and I worked together for the better part of a decade, starting when I was twenty-three at the American Council on Renewable Energy (ACORE) at the corner of 16th and K Street in Washington, DC.

A few months into the job after having spent hundreds of hours on the phone selling exhibit booths and sponsorships for the "PGRE" event—otherwise known as PowerGen

Renewable Energy—I found myself in Las Vegas organizing the event alongside my team at ACORE.

After pulling together ACORE's ten-by-ten booth on site, I was first introduced to Bill when he approached our booth during exhibit floor set up. He wanted the renewables industry to be understood and embraced in the national conversation and for the industry's risk-takers to be heard by a much larger audience.

He was a direct man, challenging all those around him in an attempt to understand their motivations and whether they were up for the challenge of revolutionizing the renewables industry. While at the event, Bill ceremoniously asked Richard Marks, a Los Angeles-based film producer, "Tell me, are you full of shit, Marks?" when they discussed how to bring renewable energy leadership into the spotlight.

Richard, who founded a communications and media company for energy, environment, and sustainability initiatives called Productions 1000, wasted no time proving to Bill that the media in America was ready to air the emerging stories of pioneering sustainability and corporate greening.

While directing an educational building green series for national public television (PBS), Bill showed Richard that dynamic leaders were truly coming together beyond the same old talking heads inside the Beltway to take on climate change directly.

After a successful PGRE, Bill and I found ourselves back at ACORE's offices in Washington, DC. Standing at six feet and three inches, Bill towered over me, but he was a gentle giant. Walking into the office, Bill called me over, which became a regular occurrence.

He sat me down to talk about something called "biomass" and how it had the power to change the world. Little did I know that introductory conversation would spark a decade-long mentorship and friendship filled with invaluable lessons for both my personal and professional life.

Sitting across the office from each other, we interacted daily on various discussions and programming around biomass and biofuels at ACORE.

Bill had a jean vest that was given to him by former President Ronald Reagan, adorned with an assortment of badges and pins speaking to biofuels, and he wore it as a daily accessory. That jean vest was a constant in my daily work life, as was Bill's constant encouragement that, "We're making impact here." He never let a day pass without challenging me.

His daily dedication to opening people's eyes to the opportunities afforded by the biomass and biofuels industries—and to leave this planet a better place for his grandchildren—mesmerized me and sold me on the belief that I was working to make an impact in this world.

From questions on why biomass and biofuels weren't getting enough prime time in ACORE's programming to making me promise to incorporate the word "biomass" in my vernacular every day, Bill constantly reminded me to pause my work from time to time and to look up to see the forest from the trees. He believed to achieve a goal, you constantly had to have the goal in front of you, maintaining a fixation and focus on the goal, no matter the immediate task. To Bill, that goal was the establishment and growth of the biomass industry as a way to combat climate change.

Bill's mentorship opened by eyes to observe and study leadership traits and management styles exhibited by early pioneers in the renewables sector that I would one day want to emulate. He believed in me and encouraged me to be reflective and celebrate those traits already manifesting in me that would one day contribute to me making an impact.

He constantly demonstrated grit, saying that one needs to take risks every day, stick with your decisions once you made them, and never stop fighting like hell. His impact on my early career left a mark on me, and I wanted to commit to write this book for risk-takers like Bill to ensure their memories and hard work would not be forgotten.

This trait of being committed to your mission and following through on plans was one of the lasting life lessons he taught me.

I missed his funeral three years ago due to a business trip, a decision I wish I could take back. I owed more to Bill than that and had squandered an opportunity to show my respect and gratitude to his family.

The eighteen years since I first met Bill have been filled with the ups and downs, highs and lows of any major rollercoaster ride or any industry that is organically growing—especially one that gets a steroid shot of investment tax creds or production tax credits from the US government from time to time.

Back before there ever was a Clean Energy Leadership Institute (CELI), Women in Renewable Industries and Sustainable Energy (WRISE) or any renewables affiliations groups, we were an industry starving for talent, embracing all those who drank the Kool-Aid and wanted to jump in heads first like the entrepreneurs we all were.

According to the *Merriam-Webster Dictionary*, grit is defined as, "firmness of character; indomitable spirit." It goes on to say that to have grit means you have courage and show the strength of your character.

A person with true grit has passion and perseverance. Essayist Edwin Percy Whipple goes farther to define grit as, "The grain of character. It may generally be described as heroism materialized, spirit and will thrust into heart, brain, and backbone, so as to form part of the physical substance of the man."

Bill personified the trait of grit and how grit tied with a firm belief in a cause can provide leadership in an industry asking for it. His beginnings set the stage for a life that was punctuated by achievements and powered by his personal tenacity and grit.

He signed up for the Marines at fifteen, lying about his age to serve in World War II. Bill went through boot camp and was about to go to the Pacific when he was found out and sent home. He later reenlisted, after working nights in a pulp mill while attending high school during the day and graduated from the US Naval Academy in Annapolis in 1951. After graduation, he was shipped off to battle in the Korean War.

I'd known Bill for several years when toward the end of my tenure at ACORE I found out that he was one of the most decorated soldiers in the Korean War, coming home with a Navy Cross, the service's highest decoration for valor after the Medal of Honor, along with the Silver Cross.

In reading reflections written about Bill after his death, Joanna Campe, Founder and Executive Director of Renew the Earth (RTE), wrote it was something, "He personally never mentioned or spoke about, but it was consistent with the extraordinary dedication and integrity that I came to admire in him. In fact, I would say that Bill was a person whose integrity was absolutely peerless, and it was through Bill that I was introduced to so many other highly dedicated colleagues at the EPA, at the State Department, and in Congress."

His Navy Cross citation stated that he engaged in, "A fierce hand-to-hand battle while under an intense concentration of hostile mortar, machine-gun, and small-arms fire. Although severely wounded during the engagement, he refused to be evacuated and, while receiving first aid, continued to issue orders and to direct the offensive operations of his unit."

In an interview with *The Washington Post*, Bill's wife, Anne Ruthling Holmberg, recounts that much of Bill's unit was killed. Her husband, suffering severe stomach wounds that he thought would be fatal, grabbed two Korean prisoners, put one of them under each shoulder, held grenades to their heads and forced them to carry him back to Allied lines.

"When he got there," she adds, "he was triaged into the group that could not be saved. Fortunately, a doctor recognized him and took him into the surgery tent and saved his life."

Following his tour in Korea, he healed from his wounds in the early 1950s, taking the opportunity to learn Russian at a language school in Germany. With that skill set, he was then able to serve on military missions to Moscow and Budapest in the 1960s.

He later was a Marine Corps aide in the late 1960s to two chiefs of naval operations and served in combat assignments during the Vietnam War. After his military retirement in 1970, he joined the Environmental Protection Agency and later spent the years between 1979 and 1990 directing the US Department of Energy (DOE)'s Office of Alcohol Fuel, where he began championing ethanol as a sustainable, alternative energy source.

Reflecting on that time period in a blog titled "The Godfather of Ethanol," which was organized by renewable fuels giant POET, Dave Vander Griend, president and CEO of

ICM, Inc., states Bill was "definitely one of the early pioneers who promoted ethanol within the beltway of DC. He was the lone voice out there back then."

While at DOE, many have stories of Bill's grit and his uncanny ability to challenge the establishment. The most well-aged one came to mind during a conversation I had with Laura Kimes, Bill's former right hand at ACORE's Biomass Coordinating Council (BCC). Going back to the Reagan Administration, this memory had been forgotten a long time ago before Laura rekindled it.

Right before the planned transition from the Carter Administration, an internal emergency meeting revealed that budget was being pulled out of the Office of Alcohol Fuel at DOE. With that, years of work cataloguing the beginnings of the modern biofuels industry were to go up in smoke along with the first concerted effort to create a rolodex of all biofuels contacts and efforts throughout the United States.

Bill, harking back to his reconnaissance days was said to have gathered all the data under the cover of night and walked away with that information, which was—and still is—illegal, only to then turn it over to government officials and the leaders of the initial biofuels trade associations after things cleared.

Grit-filled moves like these were one of the many reasons Sen. Ben Nelson (R-Nebraska) decided to hire Bill in the early 1980s to work on his sustainability efforts. During his spare time, Bill managed associations promoting solar and wind energy initiatives as well as legislative support for environmental measures.

He was a founding member of the Sustainable Energy Coalition, providing advice to the Senate and House

Renewable Energy and Energy Efficiency Caucus. He was also a key member of the organizing committee of the Environmental and Clean Energy Inaugural Balls since 1989.

During his time, many of the current CEOs in the US biofuels industry were Bill's interns, and they remained trusted advisors following his time on the Hill. Even years after, these CEOs would change their schedules or cancel calls to make time to speak with Bill any time he called to discuss his newest idea on how to "save the world."

"He was a dyed-in-the-wool American, made-in-America kind of guy," reflects Dave Vander Griend. "He lived and breathed renewable fuels. He was out there, always checking out the opportunities." Bill inspired Dave and his colleagues to do better, and they were open to that mission. This same challenge brought us together at ACORE.

When I met him back in 2004, Bill was one of those seventy-three-year-olds who still jogged at least a few miles a day. He was always in the office before I arrived at 8:50 a.m. alongside our Director of Finance, Jan Siler, and was one of the last to leave in the evenings at 8 p.m.

Bill was a cofounder of ACORE and chairman of the Biomass Coordinating Council (BCC), which brought together fifty or so biomass and biofuels companies. He called the industries the "Powerful Engine that Will" focusing on how bio-based industries would improve national, energy, and homeland security, create thousands of new basic industries and millions of jobs, and add at least 10 percent to US refining capacity.

All this led to what all of us at ACORE knew as Bill's famous "biomass wheel." Instead of an oak tree, Bill opted for a wheel. The wheel, with spokes being added by Bill faster than I could design them for BCC marketing, all focused on the importance bioenergy played as connecting a variety of social, economic, and security issues.

Bill positioned the biomass and biofuels industries at the center of the wheel. When they were deployed as mainstream industries, that would cause further effects, which Bill positioned as the spokes of the wheel. Those spokes included reduction of the trade deficit, improvement of public health, reduction of GHG emissions, and new bridges between urban and rural communities.

He was a staunch advocate of remineralization—the process that promotes the regeneration of soils and forests worldwide with finely ground rock dust as a sustainable alternative to chemical fertilizers and pesticides—as crucial to making biofuels sustainable.

Not a day went by in the late 2000s that his calendar wasn't packed with an illustrious rolodex of leaders coming in and out of ACORE's office, ranging from R. James Woolsey, former head of the CIA, and Jeff Broin, the CEO of POET (the largest biofuels company at the time), to Dave Hallberg, the first president and CEO of the Renewable Fuels Association (RFA). Bill was also never without a project, always juggling several initiatives.

Every year since, Bill spearheaded the Sustainable Energy Expo on Capitol Hill at the US House of Representatives, working closely with Carol and Jack Werner and the Environmental and Energy Study Institute (EESI) to directly educate policy-makers on renewables. Renewables back in the 2000s were still considered by many on the Hill as unattainable both from a technology production and from a financial viability perspective. He also founded and organized the Environmental Inaugural Ball every four years, celebrating new Administrations and their commitment to renewables and environmental issues.

If these events were not enough on top of his daily job wrangling biomass and biofuels leaders, his grit demonstrated itself over and over through his bullish insistence on the need to unite the Americas on a common commitment to renewables.

In 2008, Bill decided to connect with Joanna Campo, who was working with Latin American leaders on the crucial role of soil remineralization through her nonprofit, and with Carlos St. James, who was committed to developing the clean energy sector in Argentina. Together, they helped to found and organize the Latin American and Caribbean Council on Renewable Energy (LAC-CORE).

Gathering the support of Jose Maria Figueres—the former president of Costa Rica, who had defined his presidency around the cause of sustainability—they together launched the Renewable Energy Finance Forum—Latin America (REFF-LAC), to garner financiers' interest in renewables and sustainability efforts throughout Latin America. Getting financiers and technologies together helped spark the birth of the modern renewables industry throughout the Latin American region.

In recollecting those years with Bill over a separate conversation with both Laura Kimes and another colleague in the industry, Allison Archambault, we remembered numerous lunches at the Army Navy Club off Farragut Square in DC, where Bill invited us for his famous navy bean soup with iced tea lunches.

Anyone who knew Bill knew they should never turn down these lunch invitations. Laura said it best when describing those lunches as opportunities to "watch Bill create excitement and motivation and rally deep purpose among friends and new acquaintances. He had a way about him that would draw in anyone to hear what he had to say, and then they would leave with a new mission, assigned to them by Bill. Subsequent lunches would be progress reports and new missions."

"He made me believe anything is possible," recollects Larry Pearce, Executive Director of the Governors' Biofuels Coalition, when thinking back to lunch conversations over the years.

These lunches during the mid-2000s usually brought together four to five folks charged with new missions, who Bill selected as demonstrating grit. They usually included the likes of Scott Sklar, Doug Durante (Executive Director of the Clean Fuels Coalition), and Ken Westrick, who was in the process of launching a garage start-up called 3Tier. 3Tier later became the top renewables assessment and forecasting company globally, before being acquired by Vaisala in 2013.

Another notable grit-filled individual who would often be at the table was S. David Freeman—known to many as just "Dave" or also referred to as the "Green Cowboy" in DC circles. Dave had advised three presidents and led three of the largest public utilities in the United States, which included the New York Power Authority, the Los Angeles Department of Water and Power, and the Tennessee Valley Authority (TVA).

At all three, he pioneered massive energy conservation programs and contributed to early work with solar energy and electric vehicles. Looking back at these lunches, Allison reminded me that Dave had in fact worked with Bill and Scott Sklar in helping to roll out work around CAFE—or Corporate Average Fuel Economy—standards when all three of them worked on the Hill.

"You can do better," was the grit-filled message Allison got from one of these lunch gatherings with Dave and Bill.

Having just delivered on a micro-grid energy-access project in Haiti that was powered 95 percent by solar and the remainder from diesel gas, she was expecting a different response. At the very least, she had expected a congratulations for connecting thousands of Haitians to the grid after months of blood, sweat, and tears—and at points almost running out of funding.

Instead, she was met with a mission to reach 100 percent renewables penetration, fueled by the grit Bill and Dave knew Allison had. That challenging response from both mentors helped her to achieve development of micro-grid projects today that are 100 percent powered by solar.

Leading EarthSpark International, a nonprofit organization incubating businesses that solve energy poverty, Allison rose to Dave's challenge and succeeded. The organization, whose motto is, "De-risking by Doing" is in the process of growing deployment from two to twenty-four micro-grids in Haiti over the course of the next four years from 2021 to 2025 and bringing reliable, high-quality electricity to over eighty thousand people.

In all these ways, Bill personified the very essence of grit. From his stint at DOE to his constant apostolic mission around the infamous biomass wheel, he embodied a lack of a fear of failure as well as focus on long-term goals while demonstrating endurance and follow-through.

A key part of Bill's grit was tied to having a resilience to external obstacles while also exhibiting optimism, confidence, and creativity in overcoming challenges and never giving up.

He lived his professional life as if it was a marathon and not a sprint, demonstrated by his investment of time in setting up nonprofit networks, such as the Biomass Coordinating Council and Latin American Council on Renewable Energy, which slowly over time chipped away at the barriers holding back the biomass and biofuels industries.

He understood that long-term education followed up by constant focus on issues with relevance would ultimately turn naysayers to believers in championing holistic sustainability.

In her famous TED Talk, "Grit: The Power of Passion and Perseverance," Angela Duckworth noted that while courage is hard to measure, it's directly proportional to your level of grit.

More specifically, your ability to manage fear of failure is imperative and a predictor of success. The supremely gritty are not afraid to tank but rather embrace it as part of a process. Bill understood that defeat taught valuable lessons and the vulnerability of perseverance was a requisite for high achievement.

These defining principles that drove Bill's grit, including his stories of forging relationships with whomever would be part of the "coalition of the willing" along with his success in bringing biofuels to the forefront were highlighted in the book *The Forbidden Fuel.*

His grit was reinforced again in 2001 when President George W. Bush cited Bill as an American Hero. This was followed by Senator Tom Daschle (D-South Dakota), then majority leader, praising Colonel Holmberg in the Congressional Record, stating:

> *I have known Bill Holmberg ever since I came to Washington as a freshman Congressman more than twenty years ago. I know Bill*

not as a war hero, but as an indefatigable cham-
pion of the environment and as a visionary who
understood the potential of renewable fuels for
improving air quality and reducing our depen-
dence on imported oil long before they were
accepted as a viable alternative to fossil fuels.

Bill is a true American hero who stands as
a model for us all. His selfless commitment to
making the world a better place to live has been
demonstrated not only on distant battlefields,
but also by his daily pursuit of a more secure,
environmentally sustainable and just society.

Through his policymaking, his years spent at the EPA, and while walking the halls of Congress wearing an ACORE pin on his jean jacket beneath his navy-blue suit, he was a tireless advocate of renewable energy and biofuels.

To this day, I still try to utter the word "biomass," whether it's when reaching out to old colleagues in the biofuels industry or speaking with former ACORE colleagues Laura Kimes or Taylor Marshall. A few things remain constant: Bill's generosity to the nonprofits he supported, his dedication to family and country, and his commitment to public service are qualities that he tied to his grit.

In one of the last emails he sent before his passing in 2016, he wrote to Joanna Campe about an idea for an inspiring project that was ahead of his time but is now needed more than ever. In one section, Bill wrote:

We need to develop a new educational pro-
cess providing somewhat equal opportunities
for the working classes and their communities
throughout the world. They will then be able
to more fairly compete with overpowering

governmental, financial, and corporate lead-
ers. Education, leadership, advancing science
and technology, and the continual forthcoming
digital revolution are now becoming the great
equalizers. We need to start with the community
colleges where the middle class is best served
and readied for a more commanding position
in a sustainable planet.

At the end of the email, the signature included the following:

I am an Alumnus of the US Naval Acad-
emy, George Washington University, the US
Army Detachment R where I specialized in the
Russian Language and Soviet Affairs during
the Cold War; and, the Marine Corps Com-
mand and Staff College where I was the top
award winner. I am a medically Retired Marine
Lt. Colonel, highly decorated. I am also a
retired Senior Executive Service government
employee—EPA, Federal Energy Office in the
White House, Federal Energy Agency, and the
Department of Energy where my case was sent
to the Department of Justice for a Grand Jury
(1st SES to go before a Grand Jury) because of
my determination to support fuel ethanol to
reduce dependence on imported oil. Jim Wool-
sey, former Director of the CIA, was my attor-
ney. We won.

Bill's tenacity, eagle-eyed focus, and grit made his risks worthwhile, and because of him, many today in the biofuels and biomass industry continue his fight, so that one day, we can all say, "We won."

PART 3

TAKING THE RISK

CHAPTER 20

The Challenge

Miguel Stilwell d'Andrade

"You have to be a long-term player and have a certain comfort with taking the risk," reflects Miguel thinking back to EDP Renewables' acquisition of Horizon Wind Energy back in 2007.

Coming full circle with this chapter, one has to acknowledge the strides taken by the early risk-takers to set up business platforms that developed into strong energy plays here in the US market.

These platforms and the market demand they would create in turn would successfully be utilized by a diverse collective of global energy industry players that continue to innovate and mold the evolution and next phase of the renewables industry.

In entering the burgeoning US market, several companies saw the strong potential of renewables' growth and had a respect for the visionaries and start-ups who had started the market and scaled it to where it was in the early 2000s.

Despite the complexities of the US market being not just one market but in fact fifty individual state markets, traditional energy companies made the decision to take a risk. Being inspired by the homegrown nature of the sector and the commitment shown by numerous leaders, who had stuck with it from the beginning, they saw the potential of the industry.

One of those companies that was attracted to the US market was Energias de Portugal (EDP). Turning back time, this mighty company had humble beginnings. It had an undeniable and aspirational determination to be a renewable energy leader.

EDP had started to look at renewables more seriously in 2002, when it acquired an interest stake in Genesa, a subsidiary of Hidrocantabrico. It then acquired full control of Hidrocantabrico in 2004, followed by an acquisition of Nuon renewables' assets in Spain in 2005 and Agrupación Eólica in 2006.

EDP concurrently had another renewables company in Portugal called Enernova, which built and owned some of the first wind farms in that country from 1993 to 2003. EDP in 2005 merged both Enernova and Genesa into a new renewables subsidiary called Neo Energia. The new subsidiary started to gain critical mass, garnering projects outside of Iberia in France and Belgium.

EDP at that point realized, after looking at their new renewables business in terms of megawatts, that they were one of the top five renewable energy players globally. Their eyes were now set on a new market that their development team had alerted them to—the US.

On January 4, 2007, a young development executive named Miguel Stilwell d'Andrade, sat down with Neo Energia's management team in Madrid, where the company was headquartered. Miguel was head of corporate development for EDP and had already made a good impression with executives by the intellectual curiosity and thoughtfulness he approached every project with.

Neo Energia was in the process of taking on a challenge of exploring the US market and had hired a wind industry veteran, Gabriel Alonso, a Spaniard located in Philadelphia, Pennsylvania, who was previously with Gamesa Energy, USA.

Gabriel deliberated with Miguel on the best way to enter the market considering either a self-generated greenfield development route or potential acquisition of a platform or company.

Neo Energia's team wanted to grow out the business organically, site by site as a way of creating more value for EDP. The corporate side, represented by Miguel, thought the strategy should be about entering the market with accelerated growth through a corporate acquisition. Gabriel chimed in with a comment, "Well, there's this great company in the US called Horizon, but I think it's too big for you guys."

This statement challenged Miguel, who on his flight back to Lisbon wondered, *How big is too big?* and decided to make a call that day to Goldman Sachs, who was running the sales process. After a brief conversation, Goldman Sachs told Miguel that if EDP were to get in a nonbinding offer quickly, they would incorporate EDP's offer into the already running process.

Miguel immediately rang up the CEO of EDP, which at that time was António Mexia, who in turn had EDP's CFO, Nuno Alves, join in on the call.

Despite not knowing anything about the US market, EDP's executive team saw potential in Horizon and saw entering the US market as a way to address increasing demand for wind, wave, and solar power as alternatives to fossil fuels.

Miguel, António and the other executives conversed about the attractive features of the move. What intrigued them was the ability to participate in the biggest global renewables market as well as accessing an extensive pipeline of projects in several stages of development, spanning fifteen different states and with a combined aggregate generating potential of over nine thousand megawatts.

Miguel convinced António and Nuno to have EDP go ahead and participate in the process while continuing to do due diligence on the US market.

Led by a dynamic figure named Michael Skelly, Horizon Wind Energy, LLC was founded in 1998 to develop, own, and operate wind farms throughout the US. It not only had a strong track record but was known in Texas as a leader in renewables, starting the careers of many of those still in the industry today and giving rise to other renewables business across the country.

The mid-2000s were a ripe period of opportunity for renewables in the US. The country was responsible for one-quarter of the world's carbon dioxide emissions and used one-quarter of the world's crude oil. President George W. Bush had called on Americans to cut their gasoline use by 20 percent over the upcoming decade and began to support renewables development in an effort to have the US become energy independent.

Based in Houston, Texas, Horizon had a national footprint with offices in Oregon, Illinois, California, Colorado, and Minnesota. The company was formerly known as Zilkha

Renewable Energy, LLC until August of 2005 when it changed its name to Horizon Wind Energy, LLC after Goldman Sachs had bought the company from the father-son duo Selim and Michael Zilkha.

The Zilkhas were visionaries in their own right in starting the company, staying away from the press and spotlight in doing so, and focusing on their vision of building a wind powerhouse. However, in a time when most Texan energy giants did not regard renewable energy as a profitable business, Michael Zilkha was the motivation behind the company entering into the renewables business.

Via the company's former website, he stated, "There are businesses that offer the potential for greater monetary rewards. Few, however, combine the possibility of profit with societal improvement on as fundamental a level." Getting into renewables was a challenge the Zilkhas gladly accepted, and that commitment and spirit continued through the evolution of the company.

Challenges like those embarked upon by the Zilkhas are approached either with enthusiasm or dread. They usually are the epitome of decision-making in that a simple but crucial choice must be made.

Taking on or fleeing a challenge comes with an opportunity cost, which is uniquely derived by the individual making the decision, assigning a positive or negative twist to what is gained or lost by the decision.

Career Coach Marty Nemko in a *Psychologist Today* blog states that the very exercise of accessing opportunity cost can "yield wiser decisions" when weighing pros and cons of any decision in taking on a challenge.

He goes on to say that in undergoing the exercise, you evaluate your track record, lessons learned, situational locale, and the amount of effort you'd want to put into taking on the challenge. Given this, weighing opportunity costs gives one an advantage, which skews the probability of successful growth of the individual in their career, or the successful acquisition of a goal, in a positive direction. This exercise ultimately defines whether one takes a challenge, and indeed takes on the risk associated with that challenge, as well as the rewards from taking that risk.

Sitting down in Houston that February of 2007, Miguel sat face-to-face with Michael Skelly, alongside colleagues from Citigroup and Skadden Arps who were involved in the transaction from EDP's side.

They liked the Horizon team and respected what they stood for. Performing a quick due diligence around the complex transaction due to all the tax equity involved, how to account for the PTC and much more, Miguel and the team took the offer through EDP's supervisory board and put in an offer in March.

"This was incredibly quick, particularly given the size of the transaction and given the risk because of the completely different market along with the complexity around counting of the PTC and understanding the overall dynamics of the market," recollects Miguel.

They signed paperwork in March of 2007 with regulatory approval coming that June. After some time, EDP decided to merge Neo Energia, Horizon, and its other renewables business units into a new umbrella entity called EDP Renewables (EDPR) later that year. Sensing the potential to grow out the renewables business further, EDPR went through an IPO process in June of 2008, raising over $2.4 billion.

In September of 2008, the world crashed with the financial crisis.

"If we hadn't done the IPO in June before the crash, the company would have not been able to continue investing and scaling up renewables as we would have hoped to," notes Miguel. "We would have still invested, but our growth trajectory would have been at the medium scenario versus the high-growth scenario we've seen play out."

Like all transactions, you usually have to wait a year to be sure whether an acquisition was successful or not. To the market, the Horizon acquisition for $2.2 billion was seen as a very expensive transaction—and perhaps risky—at that time.

Today, the transaction would be seen as a no-brainer considering the company's position in the market, its extensive pipeline, strong team, track record, and asset base.

In reflecting on that time in the company's history, Miguel discloses, "You have to believe in the risk return you will have in order to be comfortable with the opportunities you take a risk on."

But at the same time, he adds, "I think courage is also essential because you need to be comfortable with the downsides."

Miguel's last point left me thinking. Though many of us head into a challenge willingly, how many of us go into that challenge weighing the downsides equally with the upsides and developing a level of comfort either way?

This trait, which is probably the hardest of all the traits discussed previously, was key not only to EDP's leaders in the Horizon transaction but to all the risk-takers mentioned before.

The ability to have comfort with the downsides of an associated challenge ultimately, over time, builds grit and resilience in one's ability to continuously take on a challenge or a risk. This comfort—or confidence—builds up the persistence needed to thrive in renewables.

Moving on from this poignant topic, our conversation was nearing its end, and we wanted to speak further about the importance of teamwork in willingly accepting a challenge. To Miguel, weighing risk around critical decision-making is critically tied to the importance of a strong team supporting you.

"The right team is vital to a company's DNA, and you must have faith that they will give you their best feedback and opinions. Based on that, you need to just go for it and take on the challenge in front of you," he professes.

When looking up the word "challenge" in the context of "taking a challenge," the *Cambridge English Dictionary* defines the verb as "an invitation to compete or take part" or "an invitation to do something difficult." In deviating from previous chapters' focus on traits, I was inspired by Miguel's story, and I fixated on this concept of being willing to take on a challenge in order to grow.

I had originally not been a big fan of challenges, and quite honestly, at times they would frustrate me. Just like many of us entering the middle of our careers, I had viewed challenges with a negative attitude.

To me, they were uncomfortable, an inconvenience leading me to compromise or usurping me from doing what I thought was the best. I'd readily wonder why it was worthwhile to take a risk if pain would potentially be involved in the short term, probably followed by failure or long-term scars to my career?

Being a creature of habit, repetition has always held a certainty and safety for me. However, a negative byproduct of repetition is the inability to allow yourself to grow or house the creativity needed to persist in becoming a better renewables professional.

Eventually, I had to ask myself, *What am I doing in renewables if I am unwilling to be a risk-taker?* What is anyone doing in this industry if they are risk-averse?

Being risk-averse comes to me from my upbringing growing up in an immigrant family. This mind-set locked you into setting out a path with clear line of sight of a goal, reconfirming every step of the way that your path was as linear as possible.

I trained my mind that failing would never be an option. I believed in hard work and fiscal responsibility, but my own creativity has been slow to mature. This book has shown that every single person's success came as a result of perseverance, including my own.

As a marketer, I know that going into a marketing or advertising campaign comes with no assurances of success since everything is dependent on how the target audience or customer responds. In renewables, though, you are working with an audience unlike any other, mainly one that ascribes itself to being not only an investor in business, but also an investor in society.

The renewables audience embodies those "boy scout qualities" that Mike Eckhart had referenced. Unlike in traditional marketing, we renewables marketers today are challenged to weave Environmental, Social, and (Corporate) Governance "ESG" benefits into everything we market and to push the boundaries of what our teams can deliver.

We in essence need to continuously recreate the oak tree we are inviting our audiences to gather at and make sure the messaging with our oak tree remains relevant. Ultimately, this is the main goal of any renewables marketer.

I knew from the start of my marketing career that I thrive in an entrepreneurial culture with enormous creativity and pioneering. I have had to learn not to be enveloped by an insecure fear.

Being able to deliver in renewables marketing comes with ferocious persistence as well as a belief in our ultimate product, which is to market and deliver on a transformed, sustainable energy grid.

In addition, I've learned you must be adaptable and challenge yourself to create renewables market messaging and campaigns that relate and speak directly to the swaths of new audiences and stakeholders entering the renewables market. These audiences span from local communities to the commercial and industrial (C&I) and municipality, university, schools, and hospital (MUSH) sectors.

Mindy Lubber, when reflecting on a *Bloomberg* piece titled "COVID-19 Hit Supply Chains Hard. Climate Shocks May Hurt More" tweeted that, "We too often take a stable climate for granted. Many companies are alarmingly unprepared to face climate shocks in the supply chain."

Her words should inspire us to look at how we can use marketing as a powerful tool to educate on a variety of topics to ensure our industry—and indeed our country—is ready to absorb the shocks caused by climate change.

We have no lack of talking points in marketing renewables today. From educating on the importance of building grid resiliency preparations into municipal sustainability planning to marketing the economic advantages associated with providing equal access to renewables—the messaging is clear.

However, I've learned that having hope, along with an undeterrable belief in the potential of renewables is vital.

Ultimately, we as marketers have to be creative and take risks. We have to adapt our messaging to new stakeholders. We have no other choice but to succeed.

If anything positive came out of COVID between the years of 2020 and 2022, it was a time to reflect, recharge, and revisit the reasons why I chose to be in renewables. I took the opportunity to reaffirm myself as a professional, reassess my personal career goals, benchmark my own personal successes, and look at challenges in a healthy new way.

Challenges are meant not to be easy and are a journey of growth, providing in many instances, opportunities to reflect on a journey filled with smaller pitfalls and victories that lead to an ultimate goal or accomplishment.

Maja Petrovic, a software engineer who hosts a blog about her experiences in the start-up sector states, "Rising to a challenge changes everything. It transforms our mind and widens our perspective."

The challenge takes one through emotions of fear, self-doubt, and insecurity and "ultimately leads to superior performance, proving to ourselves that we can do anything we set our minds to." Petrovic goes on to state that challenges provide mental rewards that inspire continued taking on of challenges through one's career.

These rewards range from developing mental toughness to conquering the fight-or-flight response as well as garnering courage, being cognizant of one's range of limits, focusing our attention, building self-confidence, and empowering our own self-freedom.

"Testing of courage manifests itself in the formation of character, initiative, self-control and responsible behavior," she notes. "Unless we test ourselves, we will never know our full potential."

Psychologist Caroline Beaton, through her *Forbes* piece, adds that in taking on any challenge, "The psychological key to overcoming challenge is to change our minds about what's around and ahead of us. Success depends on how we choose to see our environment."

Ultimately to Beaton, "The obstacle isn't the challenge. It's you *seeing* it as an obstacle. Transform its appearance in your mind and you may transpose a solution."

In much the same way, the birth of EDP Renewables North America exemplifies this lesson of accepting a challenge without being held back by seeing the obstacles ahead.

In embracing a career in the renewables industry, we all have accepted the challenge. To fully reach our potential though, we'll need to take the next risk.

CHAPTER 21

The Next Risk

When we hear the term risk, we automatically associate it with a negative presupposition, reinforced by *Merriam-Webster Dictionary's* characterization of it as doing, "Something that may result in loss and failure."

My intent in starting on the journey to write a book was to flip that connotation to a positive one, where in accepting a challenge, one was indeed taking a risk. Taking the risk in turn allows an individual to embark upon a journey of professional self-discovery, and in the process, creates a positive impact for the broader US renewables industry.

In a time and age where we're told to mitigate or hedge risk in our daily workplaces, we need to take on the challenge accepted by these early entrepreneurs in our industry who shirked the yokes of fear and took the risk.

Only when we take our knowledge, leverage it by mentorships, and juxtapose it against the traits exhibited by these early risk-takers do we realize we need to infuse more risk-taking in our daily lives. Ultimately, embracing risk leads to opportunities and innovation, which is what the US renewables energy industry is all about.

When pulling together this book, it was extremely difficult to narrow down from over one hundred executives who fit the "risk-taker" profile to the twenty-five I did. Even though I sorted these executives into chapters based on their dominant traits, what was truly extraordinary was that they all share the fifteen traits that I had prioritized as key. All these key traits combined produced a type of entrepreneur rarely found in other industries.

These traits—curiosity, persistence, being a connector, audacity, social entrepreneurship, being adventurous, having foresight, servant-leadership, dedication, being the first, adaptability, fearlessness, innovation, being enterprising, and grit—combined, show the resiliency of the leaders we have in our industry.

In an era when we forget the importance of having a "personal brand," these leaders show that despite common traits, we all have the ability to put our own touch on our approach to our careers while growing the renewables sector.

In addition, these leaders with shared traits formed their own oak trees, mobilizing our industry to come together. They showed us we should never underestimate the strength of coalition and community building.

Though we're a twenty-first century technology-focused industry, the people—and the values we hold—make the renewables market in the US truly exceptional and attractive.

The key question that remains is how do we accelerate the risk-taking process and approach in our industry?

Indeed, we are in an inflection point in asking, "What is our next challenge?" and "What are the next risks we will need to take to move us further forward?"

Risks we encounter remain continuously tethered to the booms and busts our industry faces today. Perceived market risks continue to multiply in the energy sector, making us sometimes feel like we've taken three steps forward yesterday and two steps backward today.

Topics of daily conversation today range from solar tariff uncertainty and competition for rare-earth minerals critical to renewable technologies to COVID-induced supply chain disruptions and Russia's invasion of Ukraine along with its effect on global energy trade.

There is no lack of external factors or events to point to in convincing ourselves we'd be better off sometimes by not taking a risk.

In times like these we must flip these negative market situations into positive ones, finding business opportunities to grow the renewables sector and capitalize on them. The profiles of those featured in this book—multiplied by hundreds if not thousands in the renewables industry across the US—serve as a reminder of how truly important it is to keep pushing the envelope.

We need to constantly assess our relevance and value proposition to the broader energy sector and solve our challenges through innovation. This will require us being conscious as to how we address and take on risks and challenges moving forward.

No matter the stage of your career in renewable energy— as someone who is newly entering the market, someone who like me is mid-career, or someone who has reached the top and is reflecting on their accomplishments—we all have the power to impact the direction of the renewables industry.

All it takes is embracing a challenge and the risk associated with it.

My hope is that we all have the opportunity to one day meet at the same oak tree, looking back at how we have empowered our world to embrace sustainability. I'm hoping through this book, I've answered your question of why we collectively need to take the risk.

We all have the courage and fortitude within us to be the very leaders our world needs, and it's all because of the industry we call our home—the US renewable energy industry.

I look forward to joining you in taking the risk.

PART 4

AFTERWORD, ACKNOWLEDGMENTS, RESOURCES AND APPENDIX

Afterword

Mona Dajani

The timing of *We Took the Risk's* publication is not a coincidence. It is clear from both an economic and environmental perspective that we have no time to lose.

Like a solar tower in a field of mirrors, Tom shines an intense beam of journalistic competence on perhaps the greatest challenge of our time while renewing a practical sense of hope that it's not too late to move toward a literally brighter future.

A master storyteller as well as a leading renewable energy expert, Tom embarks with us on a riveting story of the personal history of the early pioneers of renewable energy in the United States. *We Took the Risk* tells the inside stories, tackles the tough questions, and reveals surprising insights about the history of renewable energy.

In a sprawling story richly textured with original material, quirky details, and amusing anecdotes, he explains how climate change became a great issue and leads readers through the creation and rebirth of our country's renewable energy, energy independence, and energy security.

Epic in scope and never more timely, *We Took the Risk* vividly reveals the decisions, technologies—and most importantly—the individuals who are shaping our energy future.

This revelatory historic account of the risk-takers in renewable energy shows how those pioneers fought to make renewable energy an engine of global political and economic change.

This book contains a nice combination of interviews, stories, and examples of how our world is transitioning from a fossil fuel and nuclear infrastructure to a clean, renewable one. This important and eye-opening analysis should be read by anyone interested in emulating this feat in other countries.

I have been blessed to have known Tom for almost twenty years, when we met as young idealists who passionately believed in the mission of renewable energy to save the planet. As risk-takers, we professionally took separate paths to serve the same mission.

Still in our concentric spheres, we have traveled around the world together teaching many stakeholders the economic and moral benefits of renewable energy investments. We've worked to illustrate how governments and businesses with a far-sighted approach will reap long-term benefits while others will trail behind.

Through our successes, we have heavily relied on the business fundamentals and analysis of the economic forces that have given renewables a tailwind, all thanks to original risk-takers in the industry.

From a personal experience, I can attest that with his far-reaching historic personal insight and in-depth research,

Tom is uniquely positioned to address the present battle over climate change, which undoubtedly ranks as one of the most vital issues of our time.

Written in a straightforward and easy-to-understand conversational style, the book illuminates the risk-takers' strengths and weaknesses and humanizes them in a way that makes them all relatable. No matter the stage of career you're in, he gives us the opportunity to view the renewables industry through the eyes of these risk-takers.

The canvas of his narrative history is extensive—from the creation of ACORE and the birth of critical government policies and incentives to a synopsis of private investments in renewables. Tom's timeless book chronicles the struggle for wealth and power that has surrounded the importance for a low-carbon future. These struggles continue to fuel global rivalries, shake the world economy, and transform the destiny of men and nations alike.

Almost overnight, the United States has become the world's number-one energy powerhouse. Yet concern about energy's role in climate change is challenging the global economy and our way of life, accelerating a second energy revolution in the search for a low-carbon future.

All of this has been made starker and more urgent by the coronavirus pandemic, the pointless Russia war in the Ukraine, and the economic dark age it has wrought.

World politics is being upended as a new trade war develops between the United States and China, and the rivalry grows more dangerous with Russia, which is pivoting east toward Beijing. Vladimir Putin and China's Xi Jinping are converging both on energy and on challenging American leadership, as China projects its power and influence in all directions.

We Took the Risk is a primer for readers at all career levels on the coming energy transition and its global consequences. Tom provides a concise yet personal and comprehensive explanation for the extraordinary growth in wind and solar energy; the trajectory of the transition from fossil fuels to renewables; and the implications for industries, countries, and the climate.

Tom's book is a journal of discovery through a modernizing nation whose rapid progress was at times as startling to the author as it will be to you. Indeed, *We Took the Risk* inspires all of us to pursue renewable energy to save our planet. Our greatest fight now is over the speed to transition our energy economy.

Some of our greatest opportunities—and risks—are still ahead of us. Let's answer Tom's call to join together and collectively take a risk.

Acknowledgments

My first thanks goes to you, the reader.

Every day, hundreds of thousands of us in the renewables industry head onto construction sites and offices not knowing the impact we make on our country, fellow colleagues, team members, and business partners.

My hope for all of you is that you carve out your own renewables story. In so doing, I hope you create professional networks that lift you and infuse you with passion for an industry that truly is making a daily impact in the communities we serve around us.

This opportunity to go on this trip down memory lane for me has reinforced a greater love, respect, and true awe for the history behind the US renewable energy industry. Every executive interview I conducted for this book brought to life the real struggles renewables leaders have been through in creating the industry.

Their words conveyed the rawness of emotions, blood, and sweat expended in their undeterred mission to form our modern industry. Among one of the most common responses in reflecting on their journeys was the affirmation that they wouldn't have done things any other way.

Immense gratitude also comes to mind when looking back at my own renewables journey. We're truly fortunate as the US renewables industry to have the quantity and caliber of visionary and illustrative leaders among us that we do.

All of them persisted when so many of us would have given up. And the best part? None of them started their journeys in the same way, which is one of the most attractive features of working in renewables.

Most importantly, it does not matter how you got to the industry or your previous experience of renewables. The US industry is welcoming of all those with skill sets needed to advance the vision of our predecessors who got us to where we are today.

I'd like to thank countless individuals for inspiring and encouraging me to take on this writing project.

My first thanks goes to someone I didn't have the opportunity to give my full thanks to in person, and that is Lt. Col. William Holmberg. I embarked on the reflective journey of writing this book in his memory to celebrate the strides we all have made as a community to make renewables mainstream.

My second thanks goes to Richard Marks, a published author, fellow renewables adventurer and dear friend who encouraged me to find my voice and to tell my story from the lens of a risk-taker.

My third thanks goes to Professor Eric Koester and the team at the Creator Institute at Georgetown University. After four years of conversation, Eric successfully convinced me to agree to start this book writing journey in July of 2020.

Immense thanks goes also to the team at New Degree Press, who through countless hours of advice, coaching, and editing made a mediocre manuscript into something we all could be proud of: Asa Loewenstein, Developmental Editor;

Kat Li and Natalie Bailey on the Marketing Team; Katherine Mazoyer, Revisions and Marketing Editor; Amanda Brown, Copy Editor; Alexander Pavlovich, Layout Editor; Kyra Ann Dawkins, Head of NDP Community; Nikola Tikoski, Gjorgji Pejkovski and Bojana Gigovska on the Design Team; and Ryan Porter, Publication Strategist. Additional thanks to Colleen Daly for her copy-editing assistance.

Several beta readers also provided invaluable feedback during the last phase of my book publishing journey. Their tireless efforts assisted in making this book far stronger than I had originally anticipated. I'm forever in their debt: Alex Hobson, Katie Mehnert, Kristan Kirsh, Kristan Graf, Dr. Kristin Deason, Loretta Prencipe, Mark Riedy, Richard Marks, and Wendolyn Holland.

My journey wouldn't have been started if I hadn't had someone say yes to me and take the risk on a young twenty-year-old starting in the industry. That person was Michael Eckhart. Now retired after a successful career in the private energy sector, he founded the American Council on Renewable Energy (ACORE) and afterward rounded out his career at Citigroup.

I still remember like it was yesterday that November day following the 2004 ACORE Policy Forum when I sat across the desk from Mike and pitched him on my joining ACORE. Who knew that day was the start of a life-long passion and dedication to renewables? Thank you for taking a risk on me, Mike.

Equal thanks goes to the first person who introduced me to renewables: Jodie Roussell. Jodie, a fellow Hoya and dear friend, continues to make an impact in sustainability, working with Nestle on raising the bar on sustainable packaging.

Still further leaders taught me the traits and skill sets I needed to thrive in my career, ranging from Rob Sternthal, the founder of CohnReznick Capital and now Managing Director at Piper Sandler, to Nick Sangermano, who I met back in the days he was working at Credit Suisse who is now still in renewables working with a Family Office.

Countless colleagues joined me on my journey and through their kindness, patience, and support, we made hundreds of projects in renewables across the US possible. Spending thousands of hours together in the office, each and every one of them helped me get to where I am today, and I'd be remiss in not naming these team members:

THE ACORE ORIGINAL FOUNDERS AND TEAM MEMBERS:

Aimee Christensen, Alex Roth, Alla Lipsky, Amanda Howe, Andi Plocek, Andrew Deason, Anna Hahnemann, Austin Pugh, Brad Nordholm, Brandon Keefe, C. Boyden Gray, Caroline James, Cathy Snyder, Celeste Regan, Cheri Smith, Chris O'Brien, Christina Sookyung Jung, Christine Ervin, Cindi Eck, Dan Adler, Dan Arvizu, Dawn Butcher, Denis Hayes, Doug Berven, Emily Brown Easily, Evan Schmitt, Frank Tugwell, Griff Thompson, Gerald Borenstein, Hank Habicht, Howard Berke, Howard Learner, Ira Ehrenpreis, Jacob Susman, James Hewett, Jan Hamrin, Jan Siler, Jeff Seabright, Jing Su, Joe Desmond, John Eber, John Geesman, John Nevel, JP Jaudel, Judy Siegel, Karl Gawell, Kathy Weiss, Ken Westrick, Kevin Burns, Kevin Haley, Kevin Walsh, Kira Dickson, Kristin Deason, Laura Brenner Kimes, Lesley Hunter, Lewis Milford, Linda DaCosta, Lisa Frantzis, Manan Parikh, Maria Hilda Rivera, Mario DaSilva, Mark Anderson, Mark Milligan, Mark Riedy, Matt Cheney, Matt Ferguson,

Max Marshall, Meg Inda, Mel Jones, Michael Ware, Nancy Floyd, Nick Eisenberger, Noah Ginsberg, Professor Jeff Tester, Rachael Sherman, Risa Edelman, Rob Church, Rob Pratt, Roger Ballentine, Roger Feldman, Samantha Byrd, Scott Clausen, Scott Sklar, Skip Grow, Steve Morgan, Sue Tierney, Susan Nickey, Taylor Marshall, Thayer Thomlinson, the Honorable Dan Reicher, the Honorable Kathleen McGinty, Thomas Casten, Thomas Veague, Tracy Hall, Turner Houston, Vice Admiral Denny McGinn and Wesley Clark.

THE COHNREZNICK/COHNREZNICK CAPITAL TEAMS:
Adrian Ruiz-Grossman, Alye Villani Kelly, Andy Nguyen, Angus Chang, Anton Cohen, Brett Weal, Britta von Oesen, Chuck Ludmer, Claudine Cohen, Denis Roginskiy, Elizabeth Kaiga, Eric Van Arsdale, Evan Turner, Gary Durden, Goksenin Ozturkeri, Hamilton Foster, Ianny Ianachkova, Jeff Manning, Jeremy Swan, Joel Cohn, John Richardson, Joy Marquez, July Rogers Murawski, Justin Palmquist, Katie Garcia, Keira Lu Huang, Kristen Pranke, Lee Peterson, Luis Iglesias, Manish Hebbar, Maribeth Smialek, Mark Engstrom, Mark Taub, Melissa Joos, Michael Tatarsky, Michael Yurkerwich, Michelle Koning, Neha Mahajan, Risa Lavine, Roopak Shah, Ruthie Hunt, Ryland Perry, Shamus Ankrom, Stephanie O'Donnell, Stewart Wood, Ted Gunther, Terry Sandello and Vandana Daggumati.

THE PAST AND CURRENT RUBICON CAPITAL ADVISORS TEAM:
Aeneas Griffin, Alejandra Puente, Alexandre du Pasquier, Amaury Normand, Andrea Gilman, Andres O'Byrne, Barry Bowden, Barry O'Flynn, Bella Crane, Ben Geraghty, Carmen Lopez del Escobal, Christian Kummert, Conor Kelly,

Cormac Clay, Damianos Boulakis, Daniel Torrego, David Arron Friedman, David Fitzgerald, Eva Pietrzak, Gavin Murphy, Inidigo Rodriguez de Tembleque Echeverria, James O'Donoghue, Janice Rush, Jatin Gupta, Javier Sanz-Pastor Garcia-Escudero, Jesus Gonzalez Torrijos, Juan Felipe Tapia Chacon, Lisa Cotter, Maria Kang, Maria Lundon, Mary Coffey, Matthew Dooley, Miguel Angel Calvo Moya, Nacho Ruiz Hens, Ryan O'Hanlon, Santiago Ortiz Monasterio, Sara Shoff, Sean Diffney, Sean Fitzgerald, Sonia McMillan, Stephanie Annerose, Stephen Kearney, Will Kister and Yiming Li.

THE PAST AND CURRENT "C2" / EDPR NA DISTRIBUTED GENERATION TEAM:

Adam Woda, Ashlyn Brulato, Aye Phyu, Azim Momin, Bevan Augustine, Blair Matocha, Candice Michalowicz, Charles Roberts, Christina Davis, Christopher Rittenhouse, David Bader, David Dropkin, David Kane, David Ladeira, David Wolfert, Eduardo Pereira, Eric Jiang, Greg Rando, Gustavo Monteiro, Jackie Duckett, Jason Black, Jennifer Pettineo, Jessica Wosiack, John Kellar, John McEntire, Juli Kilpatrick, Julia McPherson, Katherine Becherer, Kendra Baldwin, Laura Wolff, Louis Langlois, Lucia Yu, Marya Friedman, Melissa Thomas, Micah Stanley, Mike Adzema, Mike Howell, Mike Kim, Mila Buckner, Nicole Haghpanah, Nicole Schwalbach, Nidgel D'souza, Nhu Tran, Nuno Goncalves, Patrick Gilmartin, PJ Vigilante, Ricardo Pereira, Rich Dovere, Rita Pereira, Ryan Marlborough, Ryan Young, Sam Egendorf, Samantha Brainard, Scott Edwards, Shea Williams, Taylor Page, Tom Cordrey, Tom LoTurco, Tim Folk, Travis Scott, Tom Starrs, Xipu Li and You-wha Kim.

Some individuals from the first moment you meet them, you know they will impact your professional and personal life forever. I'm indebted to their guidance and constant challenge to becoming a better renewables professional: Abby Hopper, Alina Zagaytova, Amy Chiang, Andrea Putman, Annika Toenniessen, Ashton Randle, Brandon Peck, Burl Haigwood, Christian Berle, Christine Church, Courtney Farr, David Samuel, Diane Leer, Elina Teplinsky, Elisa Farr, Emily Baker, Ethan Zindler, Florence Lowe-Lee, Graham Noyes, Jaime Carlson, John Mullen, Judy Goldstein, Judy Siegel, Julia Hamm, Kay McCall, Kelly Speakes-Backman, Kristen Graf, Linda Church-Ciocci, Maarten van Rossum, Mary Paul Jespersen, Matthew Leddicote, Mark Riedy, Michael Naylor, Mike Bowman, Mona Dajani, Monica Farrow, Monica Samec, Paul Adler, Pamela Sorensen, Rina Shah, Rob Corradi, Sarah Jespersen, Seth Levey, Shawn Breck, Summer Rain Ursomarso, Suzanne Hunt and the Honorable Jocelyne Croes, the former Plenopentiery Ambassador of Aruba to the United States.

Additional thanks goes to those of you who without seeing a cover or chapter believed in this book and reinforced the power we have as a community of leaders: Andrew McCormick, Aaron Lay, Aaron Lindenbaum, Abbe Kaufmann, Abby Watson, Ada Wu, Adam Ratajczak, Adekunle Awojinrin, Agustin "Gus" Abalo, Alain Halimi, Alye Villani, Alexis McLean, Amanda McGuire, Amaury Normand, Amna Khan, Amy Kurt, Cory Lankford, Andrea Al-Attar, Andrea Luecke, Andrea Michelle Farr, Andrew Ehrlickman, Andrew Kapp, Andrew Wheeler, Andy Lee, Andy Nguyen, Ann Burns, Anna Irwin, Anna Pavlova, Anna Schmitz, Anna Viggh, Anne Kershaw, Anne Nicole, Annika Toenniessen, Anthony Giuffra, Antonio Fayad, Arjun Krishna Kumar, Austin deButts, Barbara Slayton, Barbara and Ryan Wojnicki, Ben Jones, Betsy Arlen, Billy Fisher, Blair Matocha,

Boris Schubert, Brad Fierstein, Brad Goetz, Brent Stahl, Bret Kadison, Brett Levicky, Brett Weal, Brian Keane, Brian Nese, Brian Plain, Brian Powers, Brittany Lister, Brock Majewski, Bruce Murray, Chris Diaz, Carlos Mejuto, Carlton Carroll, Carol Jane Battershell, Carrie Hankey, Casey Timmerman, Catherine McLean, Chandra Brown, Chase Bice, Cheri Smith, Cheryl Comer, Chris Mathey, Christian Berle, Christian E. Ferguson, Christian Fick, Christina Brown, Christina Calabrese, Christine Ervin, Christopher Gladbach, Christopher Gooderham, Christopher Halbohn, Christopher Williams, Cindi Eck, Claire Austin, Claire Broido Johnson, Colleen Regan, Courtney Kirkland, Craig and Melissa Corica, Craig O'Connor, Daniela Pangallo, Daniela Shapiro, Danny Ptak, Darren Van't Hof, David Bader, David Burton, David Earl Mindham, David Fitzgerald, David Flynn, David Call, David Halligan, David Kane, David Palmer, David Samuel, David Silek, David South, Davion Hill, Dawn James, Deborah Keogh, Dennis Graham, Dennis Wilson, Derek Kou, Devin Carsdale, Dick Rauner, Diva Finton, Divyesh Patel, Deborah Knuckey, Donald Curry, Doug Herrema, Eddie Urbine, Elena Gerson, Elfije Lemaitre, Elina Teplinsky, Elisa Kenyon, Elizabeth Brady, Elizabeth Hur, Elizabeth Parella, Ellen Backus, Ellenjane Gonyea, Emily Baker, Emily Hughes Morilla, Eric Koester, Erik Ristow, Federico Toro, Francis Hoang, Frank Wolak, Gabriel Elsner, Gil Jenkins, Gina Walsh Cambareri, Goksenin Ozturkeri, Graham Jones, Grant Arnold, Gregory Laverriere, Hamilton Foster, Heath McMillion, Heather Graving, Heather Shelsta, Heidi Burke, Heidi Vangenderen, Henry Hely Hutchinson, Herve Billiet, Ianitza "Iany" Ianachkova, Ignacio "Nacho" Ruiz Hens, John Colleran, Jaclyn Sinclair Cicchiello, Jacob Susman, Jan Ahlen, Jan Porvaznik, Jan Siler, Jared Pilosio, Jarmila and Paul Weirich, Jason Clark, Jason

Hartke, Jason Kaminsky, Jason Stevens, Jatin Gupta, Jessica
Blond Seepersad, Jeff Bishop, Jeff Weiss, Jeffrey Mooradian,
Jenette Sears, Jennifer Argote, Jennifer and Rhys Gerholdt,
Jennifer Gillespie, Jennifer Toy Butler, Jennifer Zajac, Jeremy
Taylor, Jerry Miller, Jessica Wong, Jigar Shah, JoAnn Mil-
liken, Joe Ritter, Joel Cohn, John Eber, John Finnerty, John
M. Conti, John Marciano, John Reid, John Richardson, John
Ryan, John Sibley, John Paul Shay, Jon Farber, Jonathan Blake
Alexander, Jonathan Hartigan, Jonathan Powers, Jorge Cam-
ina, Jorge Enrique Vargas, Josè Cunningham, Joseph Cruz,
Joseph Santo, Josh Cohen, Joshua Epstein, Joshua Kunkel,
Juli Kilpatrick, Julia Hamm, Julie Randall, K Kaufmann, Kate
Bradley, Katherine Tweed, Kathleen Cully, Kathryn Garcia,
Katie Mehnert, Kay McCall, Kaylen Olwin, Kristen Deason,
Kelby Ballena, Kellie Mega, Kelly Clarke, Kelly Speakes-Back-
man, Kelsey Jae, Kevin Doffing, Kevin Gresham, Kevin Haley,
Kevin Leader, Kristen Wayne, Kristyna Giancola, Kurt
Andrews, Kyle Kennedy, Kyung-ah Park, Laura Stern, Lauren
Bartleson, Lauren Ferrante, Lauren LaSanta, Leah Loomis,
Louis Langlois, Lindsey O'Hern, Lisa Grassi, Loretta Prencipe,
Laura Lovelace, Lyra Rakusin, Magda Ferrante, Maggie Sklar,
Mandy Gunasekara, Manish Hebbar, María Hilda Rivera,
Marcelo Ortega Quintanilla, Marco Piana, Marek Hnizda,
Mark Riedy, Marko Liias, Marlene Church, Marlene Hall,
Martin Jiri Hnizda, Marwan Alaydi, Mathew Sachs, Matt Fer-
guson, Matt McMonagle, Matt Conger, Matthew Futch, Matt
Giglio, Matthew Miller, Matthew Rosenblum, Matthew Vale-
rio, Maureen Labanowski, Maureen Plain, Meagan Solomon,
Meghan Nutting, Mel Parrish Snare, Melina Acevedo, Melissa
Miller, Michael Bober, Michael Bowman, Michael Davies,
Michael Dees, Michael Finger, Michael Howell, Michael J.
Meaney, Michael Nardis, Michael Schoeck, Michael Tatarsky,

Miguel Angel Prado, Mike Casey, Mike Pariser, Michael Liebreich, Monica Samec, Nancy Burlew, Nancy Floyd, Nancy Sondag, Nat Kreamer, Nathan Lithgow, Nathan Ruiz, Nemer El Mouallem, Nhu Tran, Nicholas Tiger, Nick Addivinola, Nick Sangermano, Nina Rinnerberger, Noah Eckert, Nuno Goncalves, Orrin Cook, Pamela Genovese, Patricia "Trish" Maus, Patrick Cordova, Pernille Florin Elbech, Peter O'Neill, Rachael Terada, Raheleh Folkerts, Raina Abdi, Rasean Miller, Ray Wood, Randal Drew, Rhys Marsh, Ricardo Pereira, Rich Dovere, Richard Gaertner, Richard J. Marks, Richard Krauze, Richard Walsh, Rick Needham, Riedy Gimpelson, Rob Gramlich, Rob Church, Rob Davis, Robert Diaz-Arrastia, Robert Guerre, Robert H. Edwards Jr., Robert Scheuermann, Rob Sternthal, Roger Stark, Rory T. Huntly, Ros Runner, Russell Smith, Rutherford "Bo" Poats, Ryan Kerns, Ryland Parry, Sammy Brainard, Sandhya Ganapathy, Santosh Raikar, Sarah Bray, Sarah Eberly, Sarah Fitts, Sarah Greenberg, Scott Deatherage, Scott Good, Scott Greenberg, Scott Lusk, Scott Schieffer, Sean Carlton, Sean Lewis, Sean Rushton, Sean Gibson, Shalini Ramanathan, Shawn Rupert Breck, Shea Horner, Sheila MacKay, Sher Mathew, Shirish Jajodia, Siddharth "Sid" Tulsiani, Stephanie Baucus, Stephanie Annerose, Stephen Ezell, Steve LaForgia, Steven Munson, Sue Zoldak, Summer Rain Ursomarso, Susan and Mike Eckhart, Susan Sloan, Taylor Page, Theodore Matula, Thiam Giam, Thomas Brotanek, Thomas Purdy, Tim Buchner, Timothy Kemper, Timothy Montague, Timothy Rosenzweig, Tobin Booth, Tod O'Connor, Todd Davis, Todd Flowers, Tom Ladson, Tom Starrs, Tony Clifford, Tracy Fink, Tracy Hall, Tracey LeBeau, Twinee Madan, Victor and Melinda Guevara, Victoria Beard, Victoria Clifford, Virginia Pancoe, Vlada Epstein, Wade Williams, Wayne Muncaster, Wendolyn Holland and William Behling.

The industry was created one electron at a time. "Try to acknowledge all of those who created green electrons one by one in the past," encouraged Mary Beth Mandanas, CEO of Onyx Renewable Partners, when we were catching up on my research in the Winter of 2021.

There truly are so many I wish I could have included in this book, and I wanted to do them justice and at least mention those who come to mind who have impacted our industry in being "the firsts." From creating financial platforms and renewables technologies to social entrepreneurial ventures, they all deserve immense laud and a solid spotlight on their contribution to the US renewable energy industry:

Alexandra Liftman from the Bank of America, who collaborated with Michael Eckhart of Citigroup on the first green bonds in the US market;

David Arfin from NineDot Energy, who created the first SolarLease and then went on to build SolarCity's business model, which was then replicated and improved on by many others;

Dr. Donald Aitken, who was the Research Physicist at Stanford University in California, founder and chairman of the Department of Environmental Studies at San Jose State University in California, Executive Director of the US Department of Energy Western Regional Solar Energy Center, and Senior Staff Scientist for Renewable Energy Policy and Economics at the Union of Concerned Scientists. Dr. Aitken had served as president of the American Solar Energy Society twice and was vice president of the International

Solar Energy Society. His policy work can be attributed to helping create the first Renewable Portfolio Standard;

The EPA Green Power Partnership Team—most notably Kurt Johnson and Matt Clouse of the US Environmental Protection Agency (US EPA)—plus Jan Hamrin, Karl Rabago and Gabe Petelin of the Center for Resource Solutions, Lori Bird of the National Renewable Energy Laboratory, Ed Holt of Ed Holt and Associates, Angus Duncan and Rob Harmon from Bonneville Environmental Foundation, Jennifer Layke from World Resource Institute, Wiley Barbour and Alden Hathaway of the Environmental Resources Trust, Therrell (Sonny) Murphy and Mel Jones from Sterling Planet, Brent Alderfer and Brent Beerly from Community Energy, Dan Kalafatas and Steve McDougal from 3 Degrees—all who pioneered in the creation and implementation of the "Green Power Market" in the US, including the widespread marketing and use of Renewable Energy Credits (RECs) and direct carbon offsetting/greenhouse "GHG" protocols across the country;

Fred Potter, one of the founders alongside Doug Durante of the Clean Fuels Development Coalition and an instrumental figure in the early days of the DOE Office of Alcohol Fuels. He also, one of the founders of the Renewable Fuels Association and founded Hart Downstream Energy Services, serving as an EVP for Hart Energy;

Ira Ehrenpreis, a key venture capitalist in our space, famous for orchestrating Tesla's IPO;

Lynn Jurich, cofounder and CEO of Sunrun, a home solar power installation, financing, and leasing company that revolutionized the residential solar market in the US;

Pat Wood III, as former chair of the Texas Public Utility Commission and eventually of the Federal Energy Regulatory Commission (FERC), he helped found Texas' competitive power market and introduce competition for electric generation across the United States;

Peter Meisen, who in 1986 founded the Global Energy Network Institute (GENI), one of the first US nonprofit research educational institutes established to explore global solutions for peace and sustainable development;

Tim Mohin, who is predominantly the recognized thought leader in ESG disclosure, formerly serving as Chief Executive of the Global Reporting Initiative (GRI) and who also held sustainability leadership roles with Intel, Apple, and AMD and worked on environmental policy within the US Senate and Environmental Protection Agency; and

Tom Matzzie, Mary Beth Mandanas and the team at CleanChoice Energy who were one of the first 100 percent renewable retail electric providers in the US market.

As an industry we are very fortunate to have countless leaders who have paved the way for us. Their stories could easily create innumerable volumes of books. I wish I could have included them all in *We Took the Risk*.

In closing, I give my sincerest thanks to all those US renewables heroes among us today. Thank you for paving the way for all of us.

Resources for the Renewables Professional

Several resources written by renewables leaders are "cult classics" that I recommend you read. After going through the book writing journey now, I have a newfound deep respect for all these authors.

Ashby, Michelle. *The Modern Energy Matchmaker.* Omaha, NE: Addicus Books, Inc., 2010.

Bakke, Gretchen. *The Grid: The Fraying Wires between Americans and Our Energy Future.* London, UK: Bloomsbury USA, 2017.

Butti, Ken and Jon Perlin. *Tour of Solar History.* Cheshire, CT: Cheshire Books, 1980.

Carson, Ian and Vijay Vaitheeswaran, *Zoom: The Global Race to Fuel the Car of the Future.* New York, NY: Twelve—Hatchette Book Group USA, 2007.

Davis, Florence K., and Sarah Fitts, A.W. *Distributed Generation Law*. Chicago, IL: American Bar Association, 2020.

Doerr, John. *Speed and Scale: An Action Plan For Solving Our Climate Crisis Now*. New York, NY: Portfolio—Penguin Group (USA) Inc., 2021.

Fox-Penner, Peter. *Power after Carbon: Building a Clean, Resilient Grid*. Boston, MA: Harvard University Press, 2020.

Friedman, Thomas. *Hot, Flat, and Crowded*. New York, NY: Penguin Group (USA) Inc., 2009.

Friedman, Thomas. *The Lexus and the Olive Tree*. New York, NY: Anchor Books, 2000.

Hawken, Paul. *Regeneration: Ending the Climate Crisis in One Generation*. New York, NY: Penguin Group (USA) Inc., 2021.

Heck, Stefan and Matt Rogers. *Resource Revolution*. New York, NY: Houghton Mifflin Harcourt, 2014.

Hendricks, Bracken and Jay Inslee. *Apollo's Fire: Igniting America's Clean Energy Economy*. Washington, DC: Island Press, 2008.

Hoffman, Jane and Michael Hoffman. *Green*. New York, NY: Palgrave Macmillan, 2008.

Horn, Miriam and Fred Krupp. *Earth: The Sequel*. New York, NY: W.W. Norton and Company, Inc., 2008.

Gold, Russell. *Superpower: One Man's Quest to Transform American Energy*. New York, NY: Simon and Schuster, 2020.

Gold, Russell. *The Boom: How Fracking Ignited the American Energy Revolution and Changed the World*. New York, NY: Simon and Schuster, 2015.

Griffith, Saul. *Electrify: An Optimist's Playbook for Our Clean Energy Future*. Boston, MA: The MIT Press, 2021.

Jones, Van. *The Green Collar Economy*. New York, NY: Harper-Collins, 2008.

Kelly-Detwiler, Peter. *The Energy Switch*. Guilford, CT: Prometheus Books, 2021.

Leggett, Jeremy. *The Energy of Nations: Risk Blindness and the Road to Renaissance*. New York, NY: Routledge, 2014.

Lovins, Amory B. *Reinventing Fire*. White River Junction, VT: Chelsea Green Publishing, 2011.

Makower, Joel. *Strategies for the Green Economy: Opportunities and Challenges in the New World of Business*. New York, NY: McGraw-Hill, 2009.

Noyes, Graham. *The Carbon Rush*. Bainbridge Island, WA: Blake Island Media, 2012.

Nussey, Bill. *Freeing Energy: How Innovators Are Using Local-Scale Solar and Batteries to Disrupt the Global Energy Industry from the Outside In*. Atlanta, GA: Mountain Ambler Publishing, 2021.

Perlin, John. *Let It Shine: The 6000-Year Story of Solar Energy.* Novato, CA: New World Library, 2022.

Philips, Michael and Andrea Putman. *The Business Case for Renewable Energy: A Guide for Colleges and Universities.* APPA, NACUBO, and SCUP: Washington, DC, 2006.

Pope, Carl and Michael Bloomberg. *Climate of Hope.* New York, NY: St. Martin's Press, 2017.

Repetto, Robert. *America's Climate Problem: The Way Forward.* London, UK: Earthscan, 2011.

Rifkin, Jeremy. *The Third Industrial Revolution.* New York, NY: Palgrave Macmillan, 2011.

Scheer, Hermann. *Energy Autonomy.* London, UK: Earthscan, 2007.

Tercek, Robert. *Vaporized: Solid Strategies for Success in a Dematerialized World.* Los Angeles, CA: LifeTree Media Ltd, 2015.

The Worldwatch Institute. *State of the World: Special Focus—China and India, 2006.* New York, NY: W.W. Norton and Company Inc., 2006.

Tickell, Josh. *Biodiesel America.* Ashland, OH: Yorkshire Press, 2006.

Underwriters Laboratories Inc. *Engineering in Progress: The Revolution and Evolution of Working for a Safer World.* Oakton, VA: Ideapress Publishing, 2016.

Williams, Neville. *Chasing the Sun: Solar Adventures around the World*. Gabriola Island, BC, Canada: New Society Publishers, 2005.

Yergin, Daniel. *The Prize: The Epic Quest for Oil, Money and Power*. New York, NY: Simon & Schuster, 1990.

Zervos, Arthouros and Others. *Renewable Energy in Europe: Markets, Trends and Technologies*. London, UK: Earthscan, 2010.

Appendix

CHAPTER 1: THE PROVERBIAL OAK TREE

REN21. "What the Renewable Energy Community Thinks about the Green Recovery Plan." Accessed May 1, 2022. https://www.ren21.net/ren21-eu-green-recovery/.

CHAPTER 3: THE ROLLERCOASTER

BlackRock. "What are Megatrends?" *Themes*. Accessed February 20, 2022. https://www.blackrock.com/sg/en/investment%-2Dideas/themes/megatrends.

Brookings Institution. *Reforming Global Fossil Fuel Subsidies: How the United States Can Restart International Cooperation*. Accessed May 29, 2022. https://www.brookings.edu/research/reforming-global-fossil-fuel-subsidies-how-the-united-states-can-restart-international-cooperation/.

California Public Utilities Commission and Electricity Oversight Board. "*California's Electricity Options and Challenges.*" Report to Governor Gray Davis. August 2, 2000.

Editorial Team. "Oil Squeeze." *Time Magazine,* February 1979. https://web.archive.org/web/20091201015034/http://www.time.com/time/magazine/article/0%2C9171%2C946222%2C00.html.

Federal Energy Regulatory Commission. *"Order Directing Remedies for California Wholesale Electric Markets."* December 15, 2000.

Geman, Ben. "The Financial Meltdown's Green Aftermath." *Axios,* September 15, 2018. https://www.axios.com/2018/09/15/financial-crisis-stimulus-boost-renewable-energy.

International Energy Agency. *Net Zero by 2050.* Paris: IEA, 2021. Accessed February 15, 202 https://www.iea.org/reports/net-zero-by-2050.

Inter Technology/Solar Corporation. "White House Solar Panel." Smithsonian Institution—National Museum of American History, Washington, DC. Accessed February 17, 2022. https://americanhistory.si.edu/collections/search/object/nmah_1356218#:~:text=As%20a%20symbol%20of%20his,had%20them%20removed%20in%201986.

Mints, Paula. "Notes from the Solar Underground: The Solar Roller Coaster and Those along for the Ride—First Solar, SunPower, Q-Cells." *Renewable Energy World,* September 1, 2016. https://www.renewableenergyworld.com/solar/notes-from-the-solar-underground-the-solar-roller-coaster-and-those-along-for-the-ride-first-solar-sunpower-q-cells/#gref.

Salpukas, Agis. "70s Dreams, 90s Realities; Renewable Energy: A Luxury Now. A Necessity Later?" *New York Times,* April

11, 1995. https://www.nytimes.com/1995/04/11/business/70-s-dreams-90-s-realities-renewable-energy-a-luxury-now-a-necessity-later.html.

Sheffrin, Anjali. "What Went Wrong with California Electric Utility Deregulation?" *California Independent System Operator,* April 19, 2001. https://www.caiso.com/Documents/WhatWentWrongWithCaliforniaElectricUtilityDeregulation_AnjaliSheffrin_BerkeleyEnergyForum.pdf.

Sunrun. "Gigawatt Definition." *Solar Terms.* Accessed May 29, 2022. https://www.sunrun.com/go-solar-center/solar-terms/definition/gigawatt.

The Intergovernmental Panel on Climate Change. *Special Report: Global Warming of 1.5°C.* Geneva: IPCC, 2016. Accessed February 5, 2022. https://www.ipcc.ch/sr15/.

United Nations. *The Paris Agreement.* Paris: UNFCCC, 2015. Accessed February 1, 2022. https://unfccc.int/process-and-meetings/the-paris-agreement/the-paris-agreement.

US Energy Information Administration. *Short-Term Energy Outlook.* Washington, DC: EIA, 2022. Accessed May 11, 2022. https://www.eia.gov/outlooks/steo/report/electricity.php.

US Energy Information Administration. *Renewable Energy Consumption and Electricity Preliminary Statistics 2010.* Washington, DC: EIA, 2011. Accessed May 11, 2022. https://www.eia.gov/renewable/annual/preliminary/.

US Energy Information Administration. "Subsequent Events—California's Energy Crisis." Washington, DC: EIA: 2001. Accessed February 1, 2022. https://www.eia.gov/electricity/policies/legislation/california/subsequentevents.html#N_7_.

US Energy Information Administration. "Units and Calculators." *British Thermal Units*. Accessed May 29, 2022. https://www.eia.gov/energyexplained/units-and-calculators/british-thermal-units.php.

CHAPTER 4—RENEWABLES ARE NOT FOR EVERYONE

Global Reporting Initiative. "About GRI." *Get Started with Reporting*. Accessed January 18, 2022. https://www.globalreporting.org/about-gri/.

Maccoby, Michael. "Narcissistic Leaders: The Incredible Pros, the Inevitable Cons." *Harvard Business Review*. Accessed January 19, 2022. https://hbr.org/2004/01/narcissistic-leaders-the-incredible-pros-the-inevitable-cons.

Merriam-Webster. Online ed. s.v. "risk." Accessed January 20, 2022. https://www.merriam-webster.com/dictionary/risk.

Solar Energy Industries Association. *Solar Supply Chain Traceability Protocol*. Washington: SEIA, 2021. Accessed February 1, 2022. https://www.seia.org/research-resources/solar-supply-chain-traceability-protocol.

Vetter, David. "Just How Good an Investment Is Renewable Energy? New Study Reveals All." *Forbes*, May 28, 2020. https://www.forbes.com/sites/davidrvetter/2020/05/28/just-how-good-an-investment-is-renewable-energy-new-study-reveals-all/?sh=4200fc0f4d27.

CHAPTER 5—CURIOSITY

Kang M.J., Hsu M., Krajbich I.M., Loewenstein G., McClure S.M., Wang J.T.Y., and Camerer C.F. "The Wick in the Candle of Learning: Epistemic Curiosity Activates Reward Circuitry and Enhances Memory." *Psychol. Sci.* no. 20 (2009): 963-973.

Knowledge at Wharton Staff. "The 'Why' Behind Asking Why: The Science of Curiosity." *Wharton School of the University of Pennsylvania,* August 23, 2017. https://knowledge.wharton.upenn.edu/article/makes-us-curious/.

Loewenstein, George. "The Psychology of Curiosity: A Review and Reinterpretation." *Psychol. Bull.* Vol. 116 (1994): 75-9.

Stafford, Tom. "Why Are We So Curious?" BBC, June 2012. Accessed January 18, 2022. https://www.bbc.com/future/article/20120618-why-are-we-so-curious.

US Department of Energy. "Start Program." Accessed December 22, 2021. https://www.energy.gov/indianenergy/resources/start-program.

US Department of the Treasury. "1603 Program: Payments for Specified Energy Property in Lieu of Tax Credits." Accessed December 28, 2021. https://home.treasury.gov/policy-issues/financial-markets-financial-institutions-and-fiscal-service/1603-program-payments-for-specified-energy-property-in-lieu-of-tax-credits.

US Energy Information Administration. "Electricity Data Browsers." Accessed March 1, 2018. https://www.eia.gov/electricity/data/browser/.

Western Area Power Administration. "About WAPA." *About.* Accessed December 23, 2021. https://www.wapa.gov/About/Pages/about.aspx.

CHAPTER 6—PERSISTENCE

Deutschendorf, Harvey. "7 Habits of Highly Persistent People." *Fast Company*, April 1, 2015. https://www.fastcompany.com/3044531/7-habits-of-highly-persistent-people.

Kets de Vries, Manfred F.R. "The Fine Line Between Stubbornness and Stupidity." Leadership and Organizations (blog). *INSEAD*, October 1, 2018. https://knowledge.insead.edu/blog/insead-blog/the-fine-line-between-stubbornness-and-stupidity-10181#:~:text=Stubbornness%20makes%20us%20persevere.,-tend%20to%20be%20more%20decisive.

Schumacher Center for New Economics. "E.F. Schumacher." *People.* Accessed December 25, 2021. https://centerforneweconomics.org/people/e-f-schumacher/.

US Congress, House, Economic Recovery Tax Act of 1981, H.R. 4242, 97th Cong., 1st sess., became Public Law No: 97-34 on August 13, 1981. https://www.congress.gov/bill/97th-congress/house-bill/4242.

CHAPTER 7—CONNECTOR

Donadio, Rachel. "The Gladwell Effect." *New York Times*, February 5, 2006. https://www.nytimes.com/2006/02/05/books/review/the-gladwell-effect.html.

Gladwell, Malcolm. *The Tipping Point: How Little Things Can Make a Big Difference*. New York: Little, Brown, 2000.

Jaksch, Mary. "Success: Are you a Connector, a Maven, or a Salesman?" Talk on January 31, 2020 in New Brunswick, NJ. Rutgers University. http://water.rutgers.edu/Projects/GreenInfrastructureChampions/Talks_2020/01312020-Connector-Maven-Salesman.pdf.

New Columbia Solar. "New Columbia Solar Adds Industry Giant, Tony Clifford, to Board of Directors." *News*. Accessed December 25, 2021. https://www.newcolumbiasolar.com/new-columbia-solar-tony-clifford/.

Potter, Andrew. "A Backwards Glance at Gladwell." *MacLean's*, June 12, 2009. https://www.macleans.ca/general/a-backwards-glance-at-gladwell/.

CHAPTER 8—AUDACITY

American Council on Renewable Energy (ACORE). *Renewable Energy Communication and Policy Proposal*. Washington, DC, 2007.

American Council on Renewable Energy (ACORE). "Our Work." Accessed January 3, 2021. https://www.acore.org.

Bensmann, Martin (May 2010). "10 Years Renewable Energies Act (EEG)—Looking Back on a Success Story." *Biogas Journal,* August 5, 2016 (retrieved). http://www.kriegfischer.de/fileadmin/user_upload/news/International_EEG.pdf.

Canfield, Jim and Kraig Kamers. *CEO Tools 2.0: A System to Think, Manage, and Lead Like a CEO.* Atlanta: Aprio, 2017.

Govindarajan, Vijay. "Profitable Audacity: One Company's Success Story." *Harvard Business Review,* January 25, 2012. https://hbr.org/2012/01/profitable-audacity-one-companys-success-story.html.

Merriam-Webster. Online ed. s.v. "audacity." Accessed January 15, 2022. https://www.merriam-webster.com/dictionary/audacity.

Oxford English Dictionary. Online ed. s.v. "audacity." Accessed January 15, 2022. https://www.oed.com/.

Winrock International. "About Winrock." Accessed January 13, 2022. https://winrock.org/about/.

CHAPTER 9—SOCIAL ENTREPRENEURSHIP

CERES and IRRC. *A CERES Sustainable Governance Project Report Prepared by the Investor Responsibility Research Center.* Boston: CERES, June 2003. Accessed February 15, 2022. https://events.greenbiz.com/sites/default/files/document/CustomO16C45F42520.pdf.

Global Reporting Initiative. "About GRI." Accessed January 18, 2022. https://www.globalreporting.org/about-gri/.

Hayden, C.J. "Overcoming the Fear of Self-Promotion with C.J. Hayden." June 17, 2021. In *Serving Women on a Mission.* Produced by Caterina Rando. Podcast, MP3 audio. https://caterinarando.com/podcasts/episode-49-overcoming-the-fear-of-self-promotion-with-c-j-hayden/.

International Finance Corporation. *Who Cares Wins—Connecting Financial Markets to a Changing World.* Washington, DC: IFC, 2004. https://www.ifc.org/wps/wcm/connect/topics_ext_content/ifc_external_corporate_site/sustainability-at-ifc/publications/publications_report_whocareswins__wci__1319579355342.

Kefford, Matt. "How ESG Investing Could Solve the Challenges of Social Entrepreneurship." *BusinessBecause*, May 28, 2021. https://www.businessbecause.com/news/insights/7663/esg-investment-social-entrepreneurship.

Kell, Georg. "The Remarkable Rise of ESG." *Forbes*, July 11, 2018. https://www.forbes.com/sites/georgkell/2018/07/11/the-remarkable-rise-of-esg/?sh=7eaa89801695.

Lubber, Mindy. "Climate Change is our Generation's Greatest Challenge." Filmed May 14, 2014 in Lausanne, Vaud, Switzerland. TEDx Lake Geneva Video. https://www.youtube.com/watch?v=luNKafbFvlA&list=PLsRNoUx8w3rNmAChx4FXQFcTli7Nfi_tR&index=7.

MSCI. "The Evolution of ESG Investing." Accessed February 14, 2022. https://www.msci.com/esg-101-what-is-esg/evolution-of-esg-investing#:~:text=The%20practice%20of%20

ESG%20investing,the%20South%20African%20apartheid%20
regime.

Peek, Sean. "What Is Social Entrepreneurship? 5 Examples of Businesses with a Purpose." *CO by the US Chamber of Commerce,* July 30, 2020. https://www.uschamber.com/co/start/start-up/what-is-social-entrepreneurship#:~:text=Social%20entrepreneurship%20is%20the%20process,in%20society%20or%20the%20world.

Reed, Ken. "An Introduction to Impact Investing." *CoPeace* (blog). Accessed February 10, 2021. https://www.copeace.com/intro-to-impact-investing/.

Social Enterprise UK. "Resources for Social Enterprises." Accessed February 11, 2021. https://www.socialenterprise.org.uk/media-centre/resources-for-social-enterprises/.

Taylor, Ramon. "Philanthropists, Businesses Push for More Investment in Renewable Energy." *VOA,* January 27, 2016. https://www.voanews.com/a/philanthropists-businesses-push-greater-investment-renewable-energy/3166045.html.

TOMS. "Global Impact Report." Accessed February 14, 2022. https://www.toms.com/us/impact-report.html.

Wall Street Prep. "Largest Institutional Investors—A List of the World's 50 Largest Buy Side Firms." Accessed February 23, 2022. https://www.wallstreetprep.com/knowledge/largest-institutional-investors/.

Warby Parker. "The Whole Story Begins with You." *Buy a Pair, Give a Pair.* Accessed February 14, 2022. https://www.warbyparker.com/buy-a-pair-give-a-pair.

Warby Parker Impact Foundation. "Our Mission." Accessed February 14, 2022. https://warbyparkerfoundation.org/.

CHAPTER 10—ADVENTUROUS

CB Insights. "What is a SPAC?" Accessed February 1, 2021. https://www.cbinsights.com/research-spac-explainer.

Crosby, Harry. *Transit of Venus.* Paris, France: Black Sun Press, 1931.

Fernando, Jason. "A Guide to Initial Public Offerings (IPOs)." *Investopedia*, November 30, 2021. https://www.investopedia.com/terms/i/ipo.asp.

Geman, Ben. "Energy Industry Veterans Form SPAC to Take Climate Start-ups Public." *Axios*, September 30, 2020. https://www.axios.com/2020/09/30/climate-change-spac-ipo.

Hamberg, Jonas. "Falling Silicon Prices Shakes up Solar Manufacturing Industry." *Down To Earth (blog)*, September 19, 2011. https://www.downtoearth.org.in/news/falling-silicon-prices-shakes-up-solar-manufacturing-industry-34045.

Kanellos, Michael. "Hermann Scheer, Renewable Energy Hero, Dies." *Greentech Media*, October 15, 2010. https://www.greentechmedia.com/articles/read/herman-scheer-renewable-energy-hero-dies.

Lynum, Frank and Lars Magne Sunnana. "REC Norges største helprivate selskap." *E24*, February 2, 2007. https://e24.no/internasjonal-oekonomi/i/jdv9mw/rec-norges-stoerste-helprivate-selskap.

New York Times Dealbook Staff. "Iberdrola Renovables Completes $6 Billion I.P.O." New York Times, December 13, 2007. https://dealbook.nytimes.com/2007/12/13/iberdrola-renovables-completes-6-billion-ipo/.

Philanthropist News Digest. "Wal-Mart Heir and Philanthropist John Walton Dies in Plane Crash." Accessed February 1, 2021. https://philanthropynewsdigest.org/news/wal-mart-heir-and-philanthropist-john-walton-dies-in-plane-crash#:~:text=Wal%2DMart%20heir%20and%20philanthropist%20John%20T.,crashed%2C%20the%20Associated%20Press%20reports.

Takahashi, Dean. "2005: Hot IPO for solar cell maker SunPower." *The Mercury News*, August 18, 2014. https://www.mercurynews.com/2014/08/18/2005-hot-ipo-for-solar-cell-maker-sunpower/.

Walker, Matt. "Be Very Afraid: Uncertainty, Fear, and Achievement." *Psychology Today*, May 18, 2015. https://www.psychologytoday.com/us/blog/adventure-in-everything/201505/be-very-afraid-uncertainty-fear-and-achievement.

Werner, Tom. "Solar Power as a Clean, Reliable Solution." *Forbes*, Apr 23, 2007. https://www.forbes.com/2007/04/23/solutions-energy-werner-opinion-cz_ch_0423werner.html?sh=52359def30ef.

CHAPTER 11—FORESIGHT

Boss, Jeff. "14 Signs of An Adaptable Person." *Forbes*, September 3, 2015. https://www.forbes.com/sites/jeffboss/2015/09/03/14-signs-of-an-adaptable-person/?sh=4355bf8e16ea.

Britannica. Online ed. s.v. "transhumanism." Accessed May 29, 2022. https://www.britannica.com/topic/transhumanism.

Collins English Dictionary. Online ed. s.v. "foresight." Accessed January 24, 2022. https://www.collinsdictionary.com/us/dictionary/english/foresight.

Glasser, Maggie Wood. "4 Traits of a Highly Adaptable Person." *LinkedIn*, June 16, 2020. https://www.linkedin.com/pulse/4-traits-highly-adaptable-person-maggie-wood-glasser/.

Kelley, David James. "Dirty snow: The Impact of Urban Particulates on a Mid-Latitude Seasonal Snowpack." Thesis presented at Syracuse University, Syracuse, NY, June 2018. https://surface.syr.edu/cgi/viewcontent.cgi?article=1243&context=thesis.

Scoblic, Peter. "Learning from the Future." *Harvard Business Review*, July-August 2020 Issue. https://hbr.org/2020/07/learning-from-the-future.

Webb, Amy. "How to Do Strategic Planning Like a Futurist." *Harvard Business Review*, July 30, 2019. https://hbr.org/2019/07/how-to-do-strategic-planning-like-a-futurist.

CHAPTER 12—SERVANT LEADERSHIP

Bloomberg News Team. "GE to Buy Enron Wind Turbine Assets." *New York Times*, April 4, 2002. https://www.nytimes.com/2002/04/12/business/ge-to-buy-enron-wind-turbine-assets.html.

Dow Jones News Team. "BP Amoco Plans to Buy Remaining 50% Stake in Solarex." *New York Times*, April 7, 1999. https://www.nytimes.com/1999/04/07/business/company-news-bp-amoco-plans-to-buy-remaining-50-stake-in-solarex.html.

Evans, Will. "Profile: The Pickens Plan." *NPR*, August 12, 2008. https://www.npr.org/templates/story/story.php?storyId=93540411#:~:text=The%20Pickens%20Plan%20is%20a,gas%20as%20fuel%20for%20vehicles.

Forbes. "Profile: T. Boone Pickens." Accessed February 29, 2022. https://www.forbes.com/profile/t-boone-pickens/?sh=426f14ebf8a9.

Greenleaf, Robert K. *The Servant as Leader Pamphlet.* Westfield, IN: The Greenleaf Center for Servant Leadership, 2015.

Karsner, Andy. "Swearing in of Alexander 'Andy' Karsner, US Department of Energy (DOE) Assistant Secretary of Energy Efficiency and Renewable Energy (EERE)." US Senate Committee on Energy and Natural Resources. Streamed Live on March 23, 2006. https://www.energy.senate.gov/hearings.

Martin, Richard. "The One and Only Texas Wind Boom." *MIT Technology Review*, October 3, 2016. https://www.technologyreview.com/2016/10/03/157226/the-one-and-only-texas-wind-boom/.

Northouse, Peter. *Leadership: Theory and Practice, 6th Edition.* Thousand Oaks, CA: Sage Publications, 2013.

Spiro, Josh. "How to Become a Servant Leader." *Inc.*, August 2010. https://www.inc.com/guides/2010/08/how-to-become-a-servant-leader.html.

Williams, Neville. "Pioneers Of the Solar Industry—Peter Varadi." dasolar.com (blog). Accessed February 3, 2021. https://energy-blog.dasolar.com/blog/solar-blog/index.php/2009/08/solar-industry-pioneers-peter-varadi.

Reed, Ken. "An Introduction to Impact Investing." *CoPeace* (blog). Accessed February 10, 2021. https://www.copeace.com/intro-to-impact-investing/.

Robert K. Greenleaf Center for Servant Leadership. "What is Servant Leadership?" Accessed February 10, 2021. https://www.greenleaf.org/what-is-servant-leadership/.

US Congress, House and Senate, Energy Policy Act of 2005, 109th Congress, became Public Law 109-58 on August 8, 2005. https://www.govinfo.gov/content/pkg/PLAW-109publ58/html/PLAW-109publ58.htm.

US Department of Energy. "DOE Announces Up to $7 Million for Technology Commercialization Acceleration." DOE press release, August 29, 2008. https://www.energy.gov/articles/

doe-announces-7-million-technology-commercialization-ac-celeration.

Windpower Monthly Editorial Team. "Not Just a Marriage of Marketing Convenience, The Bonds Tying Enron and Zond into their Recent Marriage are Both Strong and True According to the New Partners." *Windpower Monthly*, June 1, 1997. https://www.windpowermonthly.com/article/958514/ not-just-marriage-marketing-convenience-bonds-tying-en-ron-zond-recent-marrige-strong-true-according-new-part-ners-share-view-ever-b.

CHAPTER 13—DEDICATION

Dargo, Sandor. "The Big Leap: Conquer Your Hidden Fear and Take Life to the Next Level by Gay Hendricks." *Sandor Dargo* (blog). Accessed January 15, 2022. https://www.sandordargo. com/blog/2021/05/22/the-big-leap-by-gay-hendricks.

Guthrie, Georgina. "How to Set Strategic Goals (With 73 Examples You Can Steal)." Accessed March 29, 2022. https://nulab.com/ blog/author/georginaguthrie/.

Huckman, Robert S. and Bradley Staats. "The Hidden Benefits of Keeping Teams Intact." *Harvard Business Review*, December 2013. https://hbr.org/2013/12/the-hidden-benefits-of-keeping-teams-intact.

Lancefield, David and Christian Rangen. "4 Actions Transforma-tional Leaders Take." *Harvard Business Review*, May 5, 2021. https://hbr.org/2021/05/4-actions-transformational-lead-ers-take.

Oxford English Dictionary. Online ed. s.v. "dedication." Accessed January 15, 2022. https://www.oed.com/.

Pintz, William. *Clean Energy from the Earth, Wind and Sun: Learning from Hawaii's Search for a Renewable Energy Strategy.* New York: Springer, 2016.

US Department of Energy. "Technology Commercialization Fund." Accessed January 16, 2022. https://www.energy.gov/technologytransitions/technology-commercialization-fund.

CHAPTER 14—BE THE FIRST

California Public Utilities Commission. "California Solar Initiative." Accessed December 3, 2021. https://www.cpuc.ca.gov/-/media/cpuc-website/files/legacyfiles/c/4221-california-solar-initiative-staff-progress-report-september-2007.pdf.

Corporate Finance Institute. "First Mover Advantage." Accessed December 6, 2021. https://corporatefinanceinstitute.com/resources/knowledge/strategy/first-mover-advantage/.

Gatti, Stefano. *Project Finance in Theory and Practice (Second Edition).* London: Elsevier, 2018.

Green Power Networking Institute (GENI). "Utility Grid-Connected Distributed Power Systems." Asheville, NC: National Solar Energy Conference (ASES), 1996. http://www.geni.org/globalenergy/library/technical-articles/links/green-power-network/ases-solar-1996/sacramento-municipal-utility-district/utility-grid-connected-distributed-power-systems/index.shtml.

Kuhne, Alana. "What is Corporate Renewable Energy Purchasing and How is it Changing?" *World Economic Forum*, October 28, 2021. https://www.weforum.org/agenda/2021/10/corporate-renewable-energy-purchasing-how-it-is-changing/#:~:text=Renewable%20energy%20purchasing%20has%20increased,mere%201.5%20GW%20in%202015.

Lynn, Bruce. "Leadership and Management / Turning Adversity to Advantage." *Bruce Lynn* (blog). Accessed December 2, 2021. https://brucelynnblog.wordpress.com/.

Powerhouse. "Generate Capital Cofounder Jigar Shah." 2021. In *Watt It Takes*. Produced by Emily Kirsch, Powerhouse. Podcast, MP3 audio. https://podcasts.apple.com/us/podcast/generate-capital-co-founder-jigar-shah/id1554962073?i=1000516025265.

Shah, Jigar. "In Memory of SunEdison Cofounder Brian Robertson." *LinkedIn*, December 22, 2015. https://www.linkedin.com/pulse/memory-sunedison-co-founder-brian-robertson-jigar-shah/.

Suarez, Fernando and Gianvito Lanzolla. "The Half-Truth of First-Mover Advantage." *Harvard Business Review*, April 2005. https://hbr.org/2005/04/the-half-truth-of-first-mover-advantage.

SunEdison LLC. "SunEdison and New Vision Technologies to Merge." SunEdison press release, December 22, 2005. https://www.prweb.com/releases/2005/12/prweb325221.htm.

Wells, Aaron. "Pioneering Solar Installation." *Boss Magazine*, June 2017. https://thebossmagazine.com/solar-installation-array-con-2/.

Yescombe, E.R. *Public-Private Partnerships—Principles of Policy and Finance*. London: Science Direct, 2007.

CHAPTER 15—ADAPTABILITY

Barth, F. Diane. "How Well Adapted Are You?" *Psychology Today*, December 23, 2017. https://www.psychologytoday.com/us/blog/the-couch/201712/how-well-adapted-are-you.

Cambridge English Dictionary. Online ed. s.v. "adaptability." Accessed January 15, 2022. https://dictionary.cambridge.org/us/dictionary/english/adaptability.

Chapman, Lizette. "Female Founders Raised Just 2% of Venture Capital Money in 2021." *Bloomberg*, January 11, 2022. https://www.bloomberg.com/news/articles/2022-01-11/women-founders-raised-just-2-of-venture-capital-money-last-year.

Delmas, Magali, Michael V. Russo, and Maria Montes-Sancho. "Deregulation and Environmental Differentiation in the Electric Utility Industry." *Strategic Management Journal*, Vol. 28, No. 2, February 2007. https://www.jstor.org/stable/20142431.

Glasser, Maggie Wood. "4 Traits of a Highly Adaptable Person." *LinkedIn*, June 16, 2020. https://www.linkedin.com/pulse/4-traits-highly-adaptable-person-maggie-wood-glasser/.

Grantz, Nele. "Why Managers Benefit from Thinking Outside the Box." *Experteer.com*, 2022. https://us.experteer.com/magazine/why-managers-benefit-from-thinking-outside-the-box/#:~:text=People%20who%20think%20outside%20the,ensure%20that%20you%20avoid%20shortsightedness.

Hannon Armstrong Capital, LLC. "Nancy Floyd | Venture Capital for Climate Tech." June 30, 2021. In *Climate Positive*. Produced by Gil Jenkins, Hannon Armstrong. Podcast, MP3 audio. https://climate-positive.simplecast.com/episodes/nancy-floyd-QWOh2a86.

CHAPTER 16—FEARLESSNESS

Mroz, Don. "How to Invigorate Innovation in a Stagnant Organization." *Wired Magazine*, 2013. https://www.wired.com/insights/2013/10/how-to-invigorate-innovation-in-a-stagnant-organization/.

Peppercorn, Susan. "How to Overcome Your Fear of Failure." *Harvard Business Review*, December 10, 2018. https://hbr.org/2018/12/how-to-overcome-your-fear-of-failure.

Tsaousides, Theo. "7 Things You Need to Know About Fear." *Psychology Today*, November 19, 2015. https://www.psychologytoday.com/us/blog/smashing-the-brainblocks/201511/7-things-you-need-know-about-fear.

CHAPTER 17—INNOVATION

Anthony, Scott. "Innovation Is a Discipline, Not a Cliché." *Harvard Business Review*, May 30, 2012. https://hbr.org/2012/05/four-in-

novation-misconceptions#:~:text=As%20The%20Little%20
Black%20Book,that%20idea%20to%20achieve%20results.

DSIRE Database. "Solar and Wind Energy Credit (Corporate) Program—Hawaii." Accessed on February 29, 2022. https://programs.dsireusa.org/system/program/detail/49/solar-and-wind-energy-credit-corporate.

Milkman, Katy and Kassie Brabaw. "Why Feeling Close to the Finish Line Makes You Push Harder." *Scientific American*, June 9, 2020. https://www.scientificamerican.com/article/why-feeling-close-to-the-finish-line-makes-you-push-harder/.

SunPower Corporation. "SunPower Signs Agreement to Acquire PowerLight Corporation." SunPower press release, November 15, 2006. https://investors.sunpower.com/news-releases/news-release-details/sunpower-signs-agreement-acquire-powerlight-corporation.

CHAPTER 18—ENTERPRISING

Brians, Paul. *Common Errors in English Usage, 3rd Edition.* Portland, OR: William, James and Company, 2013.

Consult Energy USA. "S01E03—Jeff Bishop—Key Capture Energy and Battery Storage." In *Recharge by Consult.* Produced by Will Vamplew. Podcast, MP3 audio. https://www.youtube.com/watch?v=COOFloIm9Tw.

Innio. "Company." Accessed May 10, 2022. https://www.innio.com/en/company.

CHAPTER 19—GRIT

Barnes, Bart. "William Holmberg, Decorated Marine who Became Renewable Energy Advocate, Dies at 88." *The Washington Post*, September 16, 2016. https://www.washingtonpost.com/local/obituaries/william-holmberg-decorated-marine-who-became-renewable-energy-advocate-dies-at-88/2016/09/16/854f4892-7b88-11e6-ac8e-cf8eodd91dc7_story.html.

Campe, Joanna. "Remembering Bill Holmberg—RTE Board Director (1995-2016)." *Remineralize the Earth*, December 11, 2016. https://www.remineralize.org/2016/12/remember-ing-bill-holmberg-rte-board-director-1995-2016/.

Congressional Record (Bound Edition), Volume 147: US Senate. *President Bush Recognized Lt. Col. Bill Holmberg as an American Hero*. Washington, DC: GPO, 2001. Accessed December 1, 2021. https://www.govinfo.gov/content/pkg/CRECB-2001-pt8/html/CRECB-2001-pt8-Pg11001.htm.

Farley, Janna. "The Godfather of Ethanol." *Vital by Poet*, Winter 2017 Issue. https://vitalbypoet.com/stories/godfather-of-eth-anol.

Holmberg, Mark. "My Dad Died this Week—Here's Why He's My Hero." *Richmond Times-Dispatch*, September 9, 2016. https://richmond.com/news/holmberg-my-dad-died-this-week-heres-why-hes-my-hero/article_5968778c-f87a-56f1-ba91-d334b6347515.html.

Merriam-Weber Dictionary. Online ed. s.v. "grit." Accessed December 1, 2021. https://www.merriam-webster.com/dictio-

nary/grit#:~:text=1%20%3A%20very%20small%20pieces%20
of,the%20pioneers%20survived%20the%20winter.

Whipple, Edwin Percy. *Character and Characteristic Men*. Boston:
Ticknor and Fields, 1866.

CHAPTER 20—THE CHALLENGE

Beaton, Caroline. "The Psychological Key to Beating a Challenge."
Forbes, Nov 25, 2016. https://www.forbes.com/sites/caroline-
beaton/2016/11/25/the-psychological-key-to-beating-a-chal-
lenge/?sh=13f1929438a4.

Goncalves, Sergio. "EDP to buy \$2.2 bln US Horizon Wind Energy."
Reuters, March 27, 2007. https://www.reuters.com/article/
us-edp-horizon/edp-to-buy-2-2-bln-u-s-horizon-wind-ener-
gy-idUSL2715639720070327.

Lubber, Mindy (@MindyLubber). "We Too Often Take a Stable
Climate for Granted." *Twitter,* January 10, 2022, 2:26pm. https://
twitter.com/MindyLubber/status/1480637293446709255.

Nemko, Marty. "Opportunity Cost." *Psychology Today*, November
5, 2019. https://www.psychologytoday.com/us/blog/how-do-
life/201911/opportunity-cost#:~:text=In%20making%20an%20
important%20decision,consider%20the%20pros%20and%20
cons.

Perin, Monica. "Wind Energy Visionary Flies Under the Radar."
Houston Business Journal, Apr 27, 2003. https://www.bizjour-
nals.com/houston/stories/2003/04/28/story5.html.

Petrovic, Maja. "Psychology of Challenges." Ministry of Programming—Technology (blog). Access February 28, 2022. https://medium.com/mop-developers/psychology-of-challenges-7dedc6f8a8a5#:~:text=Rising%20to%20a%20challenge%20changes,we%20set%20our%20minds%20to.

CHAPTER 21—THE NEXT RISK

Merriam-Weber Dictionary. Online ed. s.v. "take a risk." Accessed February 1, 2022. https://www.merriam-webster.com/dictionary/take%20a%20risk.

Made in the USA
Coppell, TX
23 October 2022

85170771R00190